Toward a Political
Sociology
of Science

TOWARD A POLITICAL SOCIOLOGY OF SCIENCE

Stuart S. Blume

THE FREE PRESS
A Division of Macmillan Publishing Co., Inc.
NEW YORK

Collier Macmillan Publishers
LONDON

The Free Press
A Division of Macmillan Publishing Co., Inc.
866 Third Avenue, New York, N.Y. 10022

Collier–Macmillan Canada Ltd., Toronto, Ontario

Library of Congress Catalog Card Number: 73–5291

Printed in the United States of America

printing number

1 2 3 4 5 6 7 8 9 10

Library of Congress Cataloging in Publication Data

Blume, Stuart S   1942–
    Toward a political sociology of science.

    Includes bibliographical references.
    1. Science—Social aspects. 2. Political
sociology. I. Title.
Q175.5.B58   301.5   73–5291
ISBN 0-02-904350-6

for my parents

Contents

vii

Preface

The genesis of this book was in a series of seminars on "science and the mass media" which I conducted with a group of arts and social science students at the University of Sussex. In beginning the book, my original intention was to explore what I then saw as a greatly neglected area, centering on social understanding of science: relations between scientists and journalists; treatment of science by the mass media; and general diffusion of scientific understanding. But, as time passed, I became increasingly of the opinion that, although of some interest in itself, the major sociological significance of this issue derived from the institutionalized relationships between the scientific community and the media, and from the capacity of society in general to comprehend and act upon the information received. Gradually, the framework expanded. The organized relationships of scientific community and media could be understood only in the light of a more comprehensive view of the scientific community's organization. I began to see the popularization of science as one aspect of the scientific role in modern society: one of society's many expectations of the scientist.

To be a scientist today involves far more than spending the day fiddling with complex apparatus, reading journals, and writing an occasional paper. On the one hand, there are the "externalities" of the role: the need to earn a living, for example, frequently by teaching, in order to earn research time.

The economic situation, the job market, must affect the equanimity of many individuals. But, leaving this aside, the research process itself requires the accomplished scientist to plan his work, react to mounting external pressures for "relevance" in his research, sit on committees, referee papers written by others. The growing political and economic pressures upon the scientific community are forcing scientists *qua* scientists into a greater awareness of their relations with the power system in society. This awareness has implications for both research activities of individuals and for the organization of the scientific community as a whole. Under these pressures, the "fine structure" of the scientific role is more and more apparent. Thus, the popularization activities of scientists encompass—perhaps always have encompassed—two functions. The first is educational. The second is political: an appeal for the interest, sympathy, and ultimately the support of the electorate. This second function may be seen as complementary to the advisory role of scientists within government, in the sense that the more effectively scientists can operate on the inside, the less is the need to seek resources by wider apppeal.

The social structure of science today cannot be understood without recognizing the importance of its links with the political system, and the scientific advisory role is illustrative of this fact. Modern science depends upon substantial injections of government funds. Most governments have established committees of scientists to help them better discharge their responsibilities, both toward research *per se* and toward other activities to which science may make a significant contribution. For the scientist, appointment to such committees indicates a general appreciation of the value of his work, comparable to being asked to judge scientific papers or a colleague's suitability for an academic post. The existence of these evaluative roles is a crucial aspect of the social system of science. At the same time, government science advisors are both "representing" their disciplines in the administration and (as political scientists have shown) carrying out important policy and administrative tasks.

Inexorably, I found myself driven to the view that my proper task was to try to explore the interrelationships of the scientific and political systems: to draw attention to the influence of these relationships upon the social structure of science and to the political roles which scientists are required to act out. Such a work would have to be comprehensive in scope, pointing out comparisons between the different political systems of the modern world and demonstrating the effect of political change on an historical time scale. I offer only an essay in this direction, and most of what I have to say relates to contemporary America and contemporary Britain. But if I can demonstrate the value, indeed the necessity, of a political sociology of science, in which the insights and findings of both sociologists and political scientists find a place, I may then hope that others will embark upon the greater task.

In writing this book, I have been constantly aware of the debt I owe to all with whom I have had the privilege of working: natural scientists, social scientists of all kinds, politicians, and civil servants both national and international. From all I have gained some inkling of science's multifaceted role in the modern world.

Although it is onerous to select a few individuals from a great number of valued colleagues and friends, there are three men whose advice and encouragement have meant more to me than they could have realized: Joseph Ben-David, Jack Embling, and Chris Freeman. Since at least two of them will disagree with much of what I have to say, let me hasten to absolve them from any responsibility for what I have written. Andrzej Huczynski helped with the research for chapters 5 and 8 and Dorothy Griffiths read the whole text, making useful suggestions; to both I am also very grateful.

Finally, I must thank my wife Hilary, who has lived with this book for four long years. Her tolerance and forbearance have been great—but they have never extended to obscurantism or pretentiousness. If I have avoided these traps, it is due to her influence.

I must add that all errors of fact, misunderstandings, and other failings are mine alone.

Acknowledgments

The author would like to thank the following individuals and organizations for permission to quote from copyright material:

The American Association for the Advancement of Science; The American Chemical Society; The American Jewish Committee; Miss Nancy Banks-Smith; Cambridge University Press; The Council for a Livable World; The Federation of American Scientists; Her Majesty's Stationery Office, London; The Macmillan Company; *Minerva;* The National Academy of Sciences; The National Dairy Council of England and Wales; *The New Scientist; The New Statesman;* Routledge and Kegan Paul Ltd.; The Royal Institute of Chemistry; *Science for the People;* The Scientists' Institute for Public Information; Mr. F. A. Sharman; and The Society for Social Responsibility in Science.

Toward a Political
Sociology
of Science

chapter 1
Toward a Political
Sociology of Science

INTRODUCTION

This book is founded upon the assumption that the social institution of modern science is essentially political and that, moreover, the scientific role is an integral part of the political system of the modern state. In looking at modern science, sociologists have tended to discuss it as an autonomous sub-system, insulated from social, economic, and political change in society. I do not believe that we need such a simplifying assumption. While such a conceptualization may have permitted a clearer analysis of certain aspects of scientific behavior, we are now in a position to move on—to discard this first approximation to reality. If we do so, we soon find that scientific institutions—the very condition of being a scientist—are affected by and respond to economic, social, and political change. Such adaptations may not be dissimilar from those which sociologists have shown to derive from changes in the intellectual structure of science itself.

Modern society has created a host of political functions for the scientist, ranging from advisory positions instituted by the executive in its need for scientific and technological guidance, to those allying him with the citizen in environmental and other pressure groups. In part, the range of such functions derives from the multifaceted nature of science itself: science *is* the disinterested search for new knowledge (*pace* the philosophers of science and Mertonian sociologists of science); science *is* an important source of industrial innovation (*pace* the economists); science *is* a (the?) source of socially significant and (potentially) repressive technologies (*pace* Marx, Marcuse, et al.). Scientists *are* "pure scientists" concerned with advancement of knowledge; scientists *are* "professionals" putting their expertise at the service of clients or employers; scientists *are* the high priesthood of technology-as-ideology. Their political responsibilities follow from the *fact* of being scientists—not from their other social roles.

In practice, I shall attempt to integrate the writings of political scientists into an essentially sociological framework. Political scientists have not, of course, assumed science to be independent of political change. They have been concerned largely with the interdependence of science and technology, on the one hand, and government, on the other. But the analysis has suffered from two weaknesses: it has been conducted principally from an administrative, rather than from a more broadly political, perspective; and no notion of the social/systemic nature of science has informed it. In the first part of this book, I will introduce the notion of political dependence into the sociological discussion of science; in the latter part, I will attempt to reanalyze the political science of science, and to set it into a somewhat broader framework.

SOCIAL STUDIES OF SCIENCE

Although the sociology of science is a relatively new area of inquiry, the assumptions underlying it—that science may be regarded as a social activity and that science and its social environment mutually influence one another—are far from new. In his essay "Science as a Vocation," Max Weber drew attention to some aspects of the social and intellectual situation of the academic "scientist" (a rather wider term in German). Interestingly, Weber chose to deal first with the scientist's material prospects—conditioned so powerfully by the distribution of authority within the university system—turning subsequently to issues such as the intellectual organization of science and the commitment required of the scientist.[1] Though scientific creativity has much in common with artistic creativity, Weber argued, they differ by virtue of the cumulative nature of scientific knowledge. This would have negative implications for the relations between the scientist and his vocation (why engage in an activity in which one's achievements were so

---

1 Max Weber, "Science as a Vocation," originally published in 1919, reprinted in H. H. Gerth and C. Wright Mills, (eds.) *From Max Weber* (London: Routledge, 1948), pp. 129–156.

soon rendered obsolete and irrelevant?), but positive implications in terms of the technical mastery of nature which it allowed.

In view of the remarkable growth of the scientific enterprise in the following half century, these issues have acquired a new relevance for social scientists. The material rewards of scientists—salaries—have been subjected to minute analysis by economists concerned with supply and demand in scientific manpower.[2] Are there too few or too many scientists? This is just one among many issues studied by economists interested in the "knowledge industry." Questions of the "intellectual organization" of science have long been the province of the philosophers, but sociologists have come on the scene of late, and, happily, the two modes of discourse now stimulate one another.[3] "Scientific creativity," the keystone of psychological interest in scientists,[4] also forms the basis for a sociological theory on the communal nature of science.[5] A final point in "Science as a Vocation" is of special interest in the context of the present study, although it has not given birth to any substantial research tradition. Weber enjoins his students not to expect *leadership* from scientists: "You come to us and demand from us the qualities of leadership," but this essentially political role is not what the scientist can offer, or can be expected to offer. But, in particular, World War II was to change this sense of isolation from political responsibility and political power, as "leading scientists were suddenly thrust into the limelight and emerged as public figures of almost heroic cast." [6] Although a substantial academic "politics of science" has emerged, I shall suggest later that it has adopted a very restricted perspective in comparison to the range of problems with which it might deal.

A number of specific issues attracting the attention of many groups gives some indication of the interplay of the various approaches to understanding science in its social context. Consider, for example, the possibility of the social determination of the content of science: the body of theory and hypothesis accepted and employed at any one time. Mannheim, it will be recalled, excepted the exact sciences from his theory of the social determination of knowledge, and such a view held sway among historians of science in the

---

2 See, for example, D. M. Blank and G. J. Stigler, *The Demand and Supply of Scientific Personnel* (New York: National Bureau of Economic Research, 1957); K. J. Arrow and W. M. Capron, "Dynamic Shortages and Price Rises: The Engineer-Scientist Case," *Quarterly Journal of Economics,* 73 (1959), 292.

3 Particularly as a result of the work of T. S. Kuhn, discussed below.

4 C. W. Taylor and F. Barron (eds.), *Scientific Creativity: Its Recognition and Development* (New York: Wiley, 1963).

5 Norman W. Storer, *The Social System of Science* (New York: Holt, Rinehart & Winston, 1966).

6 Lewis Coser, *Men of Ideas* (New York: A Free Press Paperback, 1970), p. 305.

1920s.[7] The prevailing "internalist" view held that advances in science were solely describable in terms of a logical development catalyzed by the efforts of (in Joseph Needham's words) "intellectual giants of mysterious origin." The intellectual applecart was somewhat upset by a paper delivered at the 1931 Congress on the History of Science and Technology by the Russian Boris Hessen, for his Marxist critique could not except science from dependence upon the socioeconomic structure. The thesis that Hessen sought to refute was a powerful one, for he chose to focus upon the work of one of the most brilliant scientists of modern times: "The phenomenon of Newton is regarded as due to the kindness of divine providence, and the mighty impulse which his work gave to the development of science and technology is regarded as the result of his personal genius." [8]

Marx, of course, had denied that history was a consequence, basically, of the deeds of great men, asserting that methods of production largely conditioned the intellectual processes of society. Hessen argued as follows. The development of manufacture, the rise of the bourgeois class and of the mercantilist philosophy [9] resulted in many technical demands, some in the realms of transportation, mining, and ballistics (warfare). Application of the applied mathematician's skills might be particularly valuable in the search for solutions to important problems in these areas. (For example, economy in transportation would result from an increase in the tonnage capacity of ships, and for this an understanding of the laws governing floating of bodies in liquids could be crucial; more efficient mining could result from better understanding of the properties of simple machines, especially the pumps used to remove water from mines. Hessen details many such problems.) Many underlying scientific problems lay in the field of mechanics, a field exercising a magnetic attraction for physicists of the seventeenth century: "We have compared the main technical and physical problems of the period with the scheme of investigations governing physics during the period we are investigating," Hessen writes, "and we come to the conclusion that the scheme of physics was mainly determined by the economic and technical tasks which the rising bourgeoisie raised to the forefront." [10] He turns then to Issac Newton's work, particularly his *Principia,* published in 1687. His case for the socioeconomic determination of New-

7   Karl Mannheim, *Ideology and Utopia* (London: Routledge Paperback, 1960).

8   Boris Hessen, "The Social and Economic Roots of Newton's 'Principia,'" in *Science at the Cross Roads* (London, 1931, reprinted by F. Cass, London, 1971, with an Introduction by P. G. Wersky), p. 151.

9   "If mercantilism was the pursuit of power by means of productivity, it was equally and perhaps more importantly the pursuit of profit by means of power." F. L. Nussbaum, *The Triumph of Science and Reason* (New York: Harper Torchbooks, 1962), pp. 66–67.

10   Hessen, "Social and Economic Roots," pp. 166–167.

ton's scientific interests is based, on the one hand, on adduction of evidence for Newton's general interest in technical matters (especially in problems of metallurgy and navigation) and, on the other, upon comparison of merchant class technical demands with specific scientific problems treated in the *Principia* (problems of floating bodies, projection of bodies through resistant media, etc.).

The impact of this revolutionary thesis was great: especially upon a small group of radical British scientists who had become interested in the history and social relations of science.[11] Hessen's conclusions fitted their own belief in the need for government support of science, as well as their political beliefs (though few joined the Communist party) rather well. He had concluded:

> Science develops out of production, and those social forms which become fetters on productive forces likewise become fetters upon science. . . . For the proletariat science is an instrument of . . . reconstruction. . . . Only such a conception of science can be its real liberator from those fetters with which it is inevitably burdened in class bourgeois society.
>
> The building of socialism not only utilizes all the achievements of human thought, but by setting science new and hitherto unknown tasks indicates new paths for its development. . . . Only in socialist society will science become the genuine possession of all mankind.

Hessen's critique provided the British scientists with a well-constructed tool for analyzing the history of science, a tool they had previously lacked.[12] Vigorously, they set to work to campaign for the promotion of science in the interests of the people. Among this group—which included Lancelot Hogben, Hyman Levy, Joseph Needham, the science writer J. G. Crowther, J. D. Bernal, and others—it was Bernal whose formulations of the views of the "social relations of science" movement were most comprehensive and rigorously documented. In the Preface to his classic and enduringly valuable work *The Social Function of Science,* Bernal writes:

> Science has ceased to be the occupation of curious gentlemen or of ingenious minds supported by wealthy patrons, and has become an industry supported by large industrial monopolies and by the State. Imperceptibly this has altered the character of science from an individual to a collective basis, and has enhanced the importance of apparatus and administration. But as these developments have proceeded in an unco-ordinated and haphazard manner, the result at the present day is a structure of appalling inefficiency both as to its internal organization and as to the means of

11 The impact has been described by a participant in the congress, J. G. Crowther. See his *Fifty Years with Science* (London: Barrie & Jenkins, 1970).

12 P. G. Wersky, Introduction to 1971 edition of *Science at the Cross Roads,* p. xviii.

application to the problems of production or welfare. If science is to be of full use to society it must first put its own house in order.[13]

Therefore, the first task was to reorganize the basic research system. Having analyzed the structure of this system in the United Kingdom and elsewhere, Bernal goes on to suggest remedies in the second part of his book. Changes would be needed in the means of recruiting scientists (and class barriers to admission would be eliminated). Laboratories would be organized differently and scientific communication rationalized (Bernal believed the scientific journal had had its day); above all, however, science would need support on a much more generous scale. How were these changes to be brought about? "Any reorganization of science must be a comprehensive task and one which cannot be undertaken alone either by scientific workers themselves or by State or economic organizations outside science, but only by all working together in an agreed direction. The question, therefore, of whether science can be reorganized at all is not simply or even principally one for scientists. It is a social and political question. Every aspect of any reorganization of science concerns the economic and political structure of society." [14]

Belief in dependence of the substance of science upon the dominant capitalist ideology was intimately bound up with the conviction that not only *should* science be reorganized, but that it *could*. British scientists who saw in the dominant ideology a restrictive or maleficent influence upon science came to believe in planning science. One, P. M. S. Blackett (later president of the Royal Society and a Nobel laureate in physics), wrote: "Unless society can use science, it must turn anti-scientific, and that means giving up the hope of the progress that is possible. This is the way that Capitalism is now taking, and it leads to Fascism. The other way is complete Socialist planning on a large scale. . . . I believe that there are only these two ways." [15] This belief in ideological determination of the structure and content of science, and the consequential view that science could and

---

13   J. D. Bernal, *The Social Function of Science* (London: Routledge & Kegan Paul, 1939), p. xiii.

14   *Ibid.,* p. 241. The mathematician Kol'man, in a paper at the 1931 congress, had spelled out a similar point in more detail, in his discussion of the state of mathematics. He wrote of the need for a plan "which can be made only in a country where the national economy and science are planned, must be drawn up on the basis of all the experience gained in scientific research institutes and works laboratories, on the basis of the requirements of the industry, agriculture and transport of the whole country . . . it must arise as the result of intensive work on the material thus collected and its foundation must be supplied by materialist dialectics." E. Kol'man in *Science at the Cross Roads,* pp. 226–227.

15   P. M. S. Blackett, *The Frustration of Science* (London: George Allen & Unwin, 1935).

should be planned, was fundamental to the perspective of early students of the social dimension of science.

Their work was a major source for the more "professional" studies which were to proliferate in postwar years: specialists in sociology, as in the other social sciences, began to share the interest of some practicing natural scientists in organization of research. What interest is there today in social (or external) influences upon scientific concepts and theories? Joseph Ben-David, in a recent survey of the scope of the sociology of science, suggested that a major difference between the literature prior to World War II and recent works has been the virtual disappearance of attempts to explain the content and theories of science on the basis of general social values.[16] Indeed, perhaps the only place for such presuppositions in the current sociology of contemporary science—and, though important, it is a limited one—is through the influence of the writings of philosopher–historian of science Thomas S. Kuhn.[17] Kuhn's views are gradually replacing more positivistic philosophies of science in the view of the scientific process favored by sociologists.[18]

Kuhn argues that scientists form closed communities with unique intellectual and normative traditions. Members of a specific (perhaps disciplinary) community are concerned essentially with the search for solutions to well-defined, consensually determined problems—rather than with refutation or establishment of major theories. In their problem-solving activities they make use of the experimental methods, theories, and general norms of conduct laid down in their own specific "tradition": this complex of norms (both moral and practical), theories, and so on, Kuhn refers to as a "paradigm." Changes in such paradigms serve to determine the boundaries between intellectual communities within science. Kuhn goes on to suggest that under normal conditions the scientific communities are insulated, in terms of their work, from the outside world, choice of research topics (problems, "puzzles") being determined by internal paradigms.[19] Occasionally, however, crises may arise within the community. Problems appear which seem to resist solution—indeed, they may inspire doubt as to the validity

16 Joseph Ben-David, "Introduction," *International Social Science Journal,* 22 (1970), 7–27.

17 Thomas S. Kuhn, *The Structure of Scientific Revolutions* (Chicago: University of Chicago Press, 1962).

18 Since in them the philosophic notions of "theory" and sociological ones of "community" reinforce one another. See, for example, J. Ben-David, "Scientific Growth: A Sociological View," *Minerva,* 2, No. 4 (1964), 455–477.

19 Mullins suggests that scientists in a particular area may be united by style of life as well as research style, instancing the camping activities of phage workers. See N. C. Mullins, "The Development of a Scientific Speciality," *Minerva,* 10, No. 1 (1972), 51–82.

of the paradigm. Under these conditions, dissatisfied scientists may cast about in the broader intellectual currents of society searching for a new approach to their specialty. The existing consensus is destroyed, as members turn hither and thither. The new paradigm, on which future consensual progress will be built, may come from anywhere, and there is a possibility that on these rare occasions external events may influence the content of science. Therefore, the place allowed for such external influences in the Kuhnian view of science is slight. Moreover, as Ben-David points out, historical studies of actual events "do not so far provide sufficient basis for sociological generalizations about the conditions under which external influences are likely to impinge on science." [20] However, though the more "externalist" views of Bernal and other prewar scientific humanists may exert little influence on professional sociologists today, they are by no means of purely historical interest.

In the first place, very similar views are held today by scientists of the left, who (like Bernal forty years ago) are dissatisfied with the "gerontocratic" leadership of science, and with the military-industrial system to whose interests alone science seems subordinate.

> Only by fundamental changes in the social and economic structure of society can the misuse of science and technology be prevented. So long as control over technology rests in the hands of corporate enterprise, and a government which functions on its behalf, scientific advance will be used to further corporate interests at the expense of the people. [21]

Scientific advances are seen as dependent upon the values of the Establishment, which excludes those "minorities" (e.g., blacks, women) with whom scientists of the left identify. Why, for example, is birth control research involved so overwhelmingly with improving the means of intrusion upon the *female* body, if not because of control of the research system by *men* in their own interest? Such views have a twofold relevance today for the student of social aspects of science. First, they suggest for our consideration problems in the social organization of science, problems apparent neither from the perspective of the outsider (however sophisticated his sociological theory) nor from that of the unconcerned scientist to whom he may turn for insight. (For example, we may choose to study women's role in the social system of science: are they discriminated against?) Second, we may ask *who* holds these views, and with what effect. What sort of scientists are they? How do they relate to the scientific community? What do they hope to achieve? These, too, are questions for the student of the social dimension of science.

20 Ben-David, "Introduction," p. 20.
21 *Science for the People* (December 1970), pp. 18–19.

In the second place, the views of Bernal and his friends on planning science (which stimulated such a substantial reaction in the 1940s) are now relatively commonplace.[22] It is widely believed that, by creating the right sort of institutional (rather than political) environment, the likelihood of making specific discoveries and inventions can be increased. By replacing professional control over scientific research with bureaucratic or organizational control, discoveries of "applied" or "economic" rather than purely scientific benefit can be promoted. Such a view commends itself to many students of "R & D management," and, above all, to those responsible for planning research and development activities at national or institutional levels. This seems to have been the view of the Soviet planners in establishing the kind of organization described (one assumes with accuracy) by Solzhenitsyn, in his novel *The First Circle:* [23]

> They first set up a special prison in 1930 . . . they wanted to see how scientists and technicians would work in prison. . . . The experiment worked very well. In normal conditions you could never have two top engineers or scientists on the same project—one would always oust the other in the fight over who was to get all the credit and the Stalin Prize. That's why all ordinary research teams consist of a colourless group around one brilliant head. But in prison it's another thing. . . . You can have a dozen academic lions living together peacefully in one den, because they've nowhere else to go. They soon get bored just playing chess or sitting around smoking. So they set about inventing something.

(In Solzhenitsyn's prison, "Mavrino," they are set to invent such devices as scrambler telephones and "voice print readers.") Whatever the contribution of wishful thinking, whatever the contribution of scientists trying to "sell" research to hardheaded managers and politicians, planners and politicians are largely sold on the idea of a connection between environment and discovery. Thus, today technocrats the world over are trying to find ways of gearing scientific discovery to social goals (e.g., health or housing improvements) or to maximization of profit. As dean Harvey Brooks pointed out, there is no point in arguing about whether science should be planned. "Science *is* planned, whether implicitly and largely as a by-product of decision-making processes external to science, or explicitly and conscious

22 The opposition to "Bernalism" was led by the "Society for Freedom in Science," founded in 1940 largely at the initiative of Professor Michael Polanyi and Dr. John D. Baker. Further references to the society and the controversy are given by Neal Wood in *Communism and British Intellectuals* (London: Gollancz, 1959), pp. 134–136. It is noteworthy that a similar society (Bund Freiheit der Wissenschaft) has recently been established in Germany.

23 Alexander Solzhenitsyn, *The First Circle* (London: Collins, Fontana Books, 1970), p. 82.

with respect to science and technology themselves." [24] Or as Lord Rothschild (head of the British government's central "think tank") has said: "It is sometimes said in justification of basic research, that chance observations made during such work, and their subsequent study may be just as important as those made during applied R & D. While there is some truth in this contention, the country's needs are not so trivial as to be left to the mercies of a form of scientific roulette, with many more than the conventional 37 numbers on which the ball may land." [25]

Since the war, governments and industrialists have put money into research in *expectation* of desired outcomes (frequently in the form of specific hardware). Specialists of new and old kinds,—systems analysts, technological forecasters, as well as economists and sociologists—have sought to inform government and industrial decision makers as to how they may increase the likelihood of achieving their expectations. The new specialists have devised new methods by which, they hope, the probability of success of various possible outcomes may be assessed, and research strategies devised in light of probabilities and desirabilities.[26] Economists have sought to find ways of assessing in advance the rate of return on future discoveries, while others (less optimistic) have sought to measure the uncertainty with which even the costs of discoveries may be assessed.[27] Finally, sociologists have tried to define the optimum values of organizational parameters affecting the rates of invention and innovation, so that these values may be maximized.[28] It is not difficult to show that, though it may no longer be

24 Harvey Brooks, "Can Science Be Planned?" in *Problems of Science Policy* (Paris: OECD, 1968), pp. 97–112.

25 Lord Rothschild, *A Framework for Government Research and Development* (London: HMSO, 1971), p. 3.

26 The literature is substantial. See, for example, Erich Jantsch, *Technological Forecasting in Perspective* (Paris: OECD, 1967); Erich Jantsch (ed.), *Perspectives of Planning* (Paris: OECD, 1969); Bertrand de Jouvenel, *The Art of Conjecture* (London, Wiedenfield & Nicholson, 1967); J. Bright (ed.), *Technological Forecasting for Industry and Government* (Englewood Cliffs, N. J.: Prentice-Hall, 1968); Robert U. Ayres, *Technological Forecasting and Long Range Planning* (New York: McGraw-Hill, 1969). See also papers by the above and by (inter alia) Marvin J. Cetron, Raoul J. Freeman, and the journal *Technological Forecasting and Social Change* (from 1969).

27 For the former, see A. V. Cohen and I. C. R. Byatt, "An Attempt to Quantify the Economic Benefits of Research" (Science Policy Studies No. 5) (London: HMSO, 1969). On the latter, see A. W. Marshall and W. H. Meckling, "Predictability of the Costs, Time, and Success of Development," in National Bureau of Economic Research, *The Rate and Direction of Inventive Activity* (Princeton: Princeton University Press, 1962), and K. P. Norris, "The Accuracy of Project Cost and Duration Estimates in Industrial R & D," *R & D Management*, 2, No. 1 (1971), 25–36.

28 See D. C. Pelz and F. M. Andrews, *Scientists in Organizations* (New York: Wiley, 1966) and other papers by Pelz (mainly in *Administrative Science Quarterly*); T. Burns and G. M. Stalker, *The Management of Innovation* (London: Tavistock, 1961,

consistent with our current sociological perspective on science to try to directly relate the content of science to ideological or structural variables, the *attempts made by others* may be of sociological interest.

The views of the planners, the technocrats, and the administrators are interesting because of the *effect they may have* upon the social structure of science. Previously, scientists retained a good deal of influence over the way science advances, in spite of their dependence upon government for necessary work facilities. As Polanyi or Kuhn would have it, the community of scientists determines the problems in basic research, in the various fields, which are most worthy of solution—in light of current paradigms. Much of the control which governments may possess, by virtue of their power over funds, is delegated to committees of scientists allowed a substantial measure of autonomy in allocating these funds. In other words, prevalent official ideologies have largely supported scientists' wishes to maintain autonomy of the scientific system, persuaded perhaps that only this could guarantee effective basic research.[29] But the current philosophies of planning are diverging from this view. To quote Lord Rothschild again: "Some people believe that eminent scientists, mathematicians and engineers should have an opportunity to express an overall view, on the nation's R & D to those ultimately in charge. But such an overall view, whether greatly desired or even ordered, is of questionable value." [30] Going on to discuss the work of the five Research Councils (collectively responsible for financing most high-quality basic research in the United Kingdom, although they have other functions), Lord Rothschild continues: "They are autonomous in respect of their programmes, an unsatisfactory situation in some cases." It is perhaps no surprise that the Rothschild proposals for improving what he regarded as an unsatisfactory situation aroused unmitigated opposition from the scientific community,[31] but were greeted with cautious approval by many economists interested in issues relating to R & D. Clearly, the gradual adoption of these planning "techniques," the gradual substitution of panels of systems analysts and economists for panels of scientists in the councils of government, will have important implications for structure of the scientific system. Because of the extent to which national scientific organization (both formal and informal) depends upon the way in which government

1966). See also C. D. Orth, J. C. Bailey, and F. W. Wolek, *Administering Research and Development* (London: Tavistock, 1965).

29 For an extreme statement of the view that only autonomy results in good research, see M. Polanyi, "The Republic of Science: Its Political and Economic Theory," *Minerva*, 1, No. 1 (1962), 54–73.

30 Rothschild, *Framework for Government Research*, p. 2.

31 Expression of this opposition is to be found in the letter columns of the newspapers (especially *The Times*) between December 1971 and February 1972, and in the pages of scientific periodicals *Nature* and *New Scientist* within the same period.

chooses to exert the influence deriving from its control of funds, perspectives on planning acquire new significance for the sociologist.

While current sociological perspectives on science find little place for the view that scientific content is socially determined, it would be wrong to conclude that the conviction that there is such a relationship is no longer relevant. Such views are interesting today mainly through the relationship of those holding them to the scientific community, whether the relationship is based upon participation (as with left-wing scientific groups) or control (as with science policy makers). Appropriately, sociologists may ask: Who holds such views, and with what effect? These questions will direct the investigator's attention toward views and activities of critics of the scientific establishment, interested in social and ideological determination of the rate and direction of invention. How widespread are such views within the scientific community? What effect have they had upon the institutions and organization of science? What do those subscribing to them hope to achieve in practical terms, and by what strategies? How do the scientific communities of different countries differ in the extent of their commitment to such ideas, and why? The same initial interest will direct the student of the sociology of science toward an interest in the exercise of authority in the scientific community,[32] and in the relationship of the scientific elite (to which power is largely delegated) to government. What effect does such a relationship exert on the scientific community itself?

In postwar years, then, the sociology of contemporary science has been little concerned with social determination of the substance of science: its expansion has been in other directions. Ben-David has suggested that the literature may be classified in terms of the kinds of variables used in analysis: he distinguishes two major approaches.[33] The *interactional approach* deals with the relationships between individual scientists: their patterns of communication and interaction in laboratory and disciplinary groups. The *institutional* approach adopts a more macroscopic perspective, considering the broader economic, social, and religious influences upon scientific organization, and upon definition of the scientist's role in society. An approach of this kind is almost inevitably comparative, whether between countries or between periods in history, or both. (Therefore, it is far less suited—if at all—to examination by modern survey techniques or to quantification: it has been little utilized by those American sociologists to whom the growth of the field owes most.) Today most work in the sociology of science is involved with interactional study of the scientific community.

32   Robert Merton has already begun to look at this. See H. Zuckerman and R. K. Merton, "Patterns of Evaluation in Science: Institutionalisation, Structure and Functions of the Referee System," *Minerva*, 9, No. 1 (1971), 66–100.

33   Ben-David, "Introduction," p. 15 *et seq.*

Studies have focused mainly upon relationships within research groups and laboratories (and their effects upon productivity [34]), communication networks within science,[35] and patterns of communication and interaction within specific areas of scientific research.[36] Surveying this work, Ben-David concludes that a recent shift in interest "from laboratory work groups to networks encompassing distinct fields of research was greatly influenced by the emergence of a view of science as the work of a community in the sociological sense," a view which he attributes largely to the work of Polanyi, Shils, and Kuhn.[37] The first detailed exploration of the norms, values, and structure of a national scientific community was that of Hagstrom.[38]

Systematic approaches to what Ben-David calls an "institutional" sociology of science are rather rare.[39] Many studies of "national science

34 See, for example, D. C. Pelz and F. M. Andrews, *Scientists in Organizations;* W. Kornhauser, *Scientists in Industry* (Berkeley: University of California Press, 1962); S. Marcson, *The Scientist in American Industry* (New York: Harper and Row Publishers, 1960); Barney G. Glaser, *Organizational Scientists: Their Professional Careers* (Indianapolis: Bobbs-Merrill, 1964); S. Cotgrove and S. Box, *Science, Industry, and Society* (London: Allen & Unwin, 1970); H. Baumgartel, "Leadership, Motivations and Attitudes in Research Laboratories," *Journal of Social Issues,* 12 (1956), 24–31; M. Abrahamson, "The Integration of Industrial Scientists," *Administrative Science Quarterly,* (1964), 208–218; N. Kaplan, "Professional Scientists in Industry: An Essay Review," *Social Problems,* 13 (1965), 88–97; G. M. Swatez, "The Social Organization of a University Laboratory," *Minerva,* 8, No. 1 (1970).

35 See, for example, Herbert Menzel, "Review of Studies in the Flow of Information among Scientists" (New York: Columbia University Bureau of Applied Social Research, mimeod, 1958); various reports of the American Psychological Association project on Information Exchange; A. E. Bayer and J. Folger, "Some Correlates of a Citation Measure of Productivity in Science," *Sociology of Education,* 39 (1966), 381–390; S. Passman, *Scientific and Technological Communication* (London: Pergamon, 1970); J. Martyn and A. Gilchrist, *An Evaluation of British Scientific Journals* (London: Aslib, 1968); A. de Reuck and J. Knight, *Communication in Science* London: Churchill, 1967); various papers by Derek Price and his *Science Since Babylon* (New Haven: Yale University Press, 1961) and *Little Science Big Science* (New York: Columbia University Press, 1963).

36 See, for example, Diana Crane, "Social Structure in a Group of Scientists: A Test of the 'Invisible College' Hypothesis," *American Sociological Review,* 34 (1969), 335–352 and "Trans-national Networks in Basic Science," *International Organization,* 25 (1970), 585; J. Gaston, "Secretiveness and Competition for Priority of Discovery in Physics," *Minerva,* 9 (1971), 4; N. C. Mullins, "'The Distribution of Social and Cultural Properties in Informal Communication Networks among Biological Scientists," *American Sociological Review* (1968), and "The Development of a Scientific Speciality," *Minerva,* 10, No. 1 (1972).

37 M. Polanyi, *The Logic of Liberty* (London: Routledge, 1951), p. 53; see also his "The Republic of Science," *Minerva,* 1, No. 1 (1962); Edward A. Shils, "Scientific Community: Thoughts after Hamburg," *Bulletin of Atomic Scientists,* 10 (May 1954), 151–155; Thomas S. Kuhn, *Structure of Scientific Revolutions.*

38 Warren O. Hagstrom, *The Scientific Community* (New York: Basic Books, 1965).

39 See, for example, Bernard Barber, *Science and the Social Order* (New York: Free Press, 1952); D. S. L. Cardwell, *The Organization of Science in England* (London: Heinemann, 1957); R. Gilpin, *France in the Age of the Scientific State* (Prince-

policies" carried out on a comparative basis by international organizations such as UNESCO and OECD are concerned with characterization of national scientific organization, but usually they lack a systematically developed analytical framework.[40] And, of course, governments themselves are concerned, on occasion, to modify those structural and other parameters believed to determine the rates of scientific discovery and innovation.[41] The exigencies of decision making, however, seem to preclude any great interest in systematic investigation of these parameters. Ben-David, in a recent book, sought first to explain the rise of the scientific role in seventeenth-century England and, subsequently (and more originally), to explain differences in levels of scientific activity among major industrialized countries.[42] He conceives of "levels of activity" in terms of such factors as extent of scientific publication in each country, numbers of foreign scientists received for study purposes into each country, and so on: Ben-David is concerned with changes in relative levels (so conceived). He introduces the notion of a "centre" (the nation preeminent in terms of scientific activity at any one time), seeking to explain the migration of this preeminence successfully from England to France to Germany and finally to the United States. What kinds of explanatory variables can be adduced in attempting to understand what amounts to the "effectiveness" of national scientific organization and institutions? The first explanatory variable used by Ben-David relates to initial establishment of the scientific role: "the changing constellation of social values and interests among the population as a whole." [43] The second, which assumes importance when scientific work has already begun to offer vocational opportunities in a country, is "the organization of scientific work which was more or less effective in marketing the products of research and encouraging initiative and efficiency in it." [44] Here, then, is how Ben-David

---

ton: Princeton University Press, 1968); Hans Skoie, "The Problems of a Small Scientific Community: the Norwegian Case," *Minerva*, 7, No. 3 (1969); and T. Dixon Long, "Politics and Policy in Japanese Science: the Persistence of a Tradition," *Minerva*, 7, No. 3; J. Ben-David, *Fundamental Research and the Universities* (Paris: OECD, 1968).

40 See the series "Studies and Documents of Science Policy" (Paris: UNESCO) and "Reviews of National Science Policy" (Paris: OECD).

41 See official reports such as, in the United Kingdom, that of the "Committee of Enquiry into the Organization of Civil Science," (London: HMSO, 1963), "A Framework of Government Research and Development" (London: HMSO, 1971); in the United States, "Science and Technology: Tools for Progress" (Report of the President's Task Force on Science Policy, 1970); in Canada, Report of the Senate Commission under Senator Lamontagne (1971).

42 J. Ben-David, *The Scientist's Role in Society* (Englewood Cliffs, N. J.: Prentice-Hall, 1972).

43 *Ibid.*, p. 169.

44 *Ibid.*, p. 169.

would have us comprehend current differences in national scientific organization: in terms of user demand for the results of science, perhaps the result of effective "marketing" by the scientific community. "After (1840) the jumps in scientific activity occurred as a result of the discovery of new uses of science leading to changes in the definition of the scientists' role; and innovation in research plant and organization." [45]

Necessarily, this brief account of the sources and tendencies in the sociology of science has been oversimplified. One objection which will occur to the reader is that perhaps it exaggerates what can only be an analytical distinction between interactional and institutional approaches to study of the social dimension of science. I have suggested that social, political, and economic conditions (particularly the latter) serve largely to determine the structure of scientific organizations in any country; and that for scientists working in the country in question both economic conditions, on the one hand, and the derived scientific organization, on the other, are matters of unquestionable fact. "Surely," the reader may ask, "organization and interaction are interrelated? At the level of the scientific institution—industrial, governmental, or academic research unit—doesn't the organization serve to determine the interactions taking place between scientists, and can't the scientists influence the organization?" Equally, one might be driven to inquire into the reciprocal effects of individual scientists *upon* the fundamental economic and political conditions. In spite of the legitimacy of these questions, they have been little examined by sociologists. Take the question of scientists' influence (in groups or singly) upon national scientific organizations, on the one hand, and upon the organization of a specific research institution, on the other. The ways in which scientists organize, and are organized, to exert influence over such issues will constitute a major theme of this book, but we must turn to the political scientist for what preliminary information is to be found. Therefore, let us examine the writings of political scientists upon the natural sciences. What issues have engaged their attention, and how do their findings relate to those of sociologists? I shall suggest that many problems in the relations between science and society, so far visible only from the political scientist's perspective, have considerable importance for the sociological understanding of science.

The writings of the prewar British "scientific humanists" (Bernal, Crowther, Needham, and others), by common consent a significant early attempt at sociological analysis of science, cannot be divorced from the political concerns behind them. One is inclined to doubt that they would have been impressed by the sharp separation which has sprung up today between the politics and the sociology of science as areas of inquiry. To take one example of an issue interesting them, Bernal's concern with the

45 *Ibid.,* pp. 170–171.

scientific career was much more eclectic than that of sociologists today: like Weber, but unlike them, he was as involved with the economic and living standards of "scientific workers" of all kinds as in the autonomy allowed those engaging in basic research to pursue problems of theoretical interest. And, of course, for Bernal the principal determinant of all aspects of the scientific career in Britain was the dominant capitalist ideology. Where today the sociologist of science may look at the strains to which an individual socialized into the value system of pure science but working in a mission-oriented industrial laboratory is subject,[46] Bernal was inclined to examine determinants of the laboratory ethos and, more, to *urge* scientists to join trade unions.[47] In that way, and only in that way, could they assert direct influence over their working conditions—over salaries as over work autonomy. Fundamental change would and could take place only in full appreciation of relations between scientific organization and the political system. "The activity we have been discussing so far is one inside the structure of science itself, but clearly it cannot remain limited in this way. In so far as the scientist, individually or corporately, is attempting to influence society, he is acting politically." [48]

In postwar years, parallel with the growth of professional sociological interest in science, an area of study has developed which may be termed "science and government." Dean Don K. Price is often recognized as father of this area of political science or public administration, and his book *Government and Science* (published in 1954) is a major early classic in the field.[49] Moreover, within the brief compass of his book, Price sketched out the problems which, still almost exclusively, have engaged the attention of political scientists interested in science and government. These include impact of the scientific attitude upon political thought and authority and upon the machinery of government; growth of a scientific capability within government; relationships between science-performing, -funding, and -contracting agencies and the Executive; legislation leading to establishment of scientific and technological agencies and capabilities (e.g., the National Science Foundation, the Atomic Energy Commission); government research and

---

46 W. Kornhauser, *Scientists in Industry: Strains and Accommodations* (Berkeley: California University Press, 1962).

47 Bernal, *Social Function of Science,* pp. 385–407.

48 *Ibid.,* p. 402.

49 Don K. Price, *Government and Science* (New York: Oxford University Press, 1962). In the Preface, Price writes:

I began to be aware that the activities of scientists, which had always been unusually influential in the public policies of the United States, were becoming responsible for significant changes in the nature of the American governmental system. The subject seemed to me to cry out for attention and to involve a whole series of most profound and most neglected questions (p. vi).

development contracting; erosion of the integrity of science in the light of national security considerations, and locus of control over application of science and technology in public policy; and establishment and operation of advisory machinery within government. In the main, political scientists (and a few historians) have tended to interest themselves in just these problems: the growth of scientific capability within government; [50] legislation; [51] government research and development contracting and program management; [52] and the personalities involved in, and operation of, the scientific advisory machinery.[53] Relatively few authors have attempted to confront problems less directly related to the *administration* of science.[54] Attempts to assess the impact of political institutions and political ideology *upon* science have been limited largely to studies of the aberrations of Stalin's Russia.[55]

In recent years, the development of the subject must have been affected

50 See, for example, A. H. Dupree, *Science in the Federal Government: A History of Policies and Activities to 1940* (Cambridge: Harvard University Press, 1957); Robert R. Gilpin, *France in the Age of the Scientific State* (Princeton: Princeton University Press, 1969); Dorothy Schaffter, *The National Science Foundation* (New York: Praeger, 1969); Eugene Skolnikoff, *Science, Technology and American Foreign Policy* (Cambridge: MIT Press, 1967).

51 See, for example, Norman Vig, *Science and Technology in British Politics* (London: Pergamon, 1967); J. S. Dupre and S. A. Lakoff, *Science and the Nation* (Englewood Cliffs, N. J.: Prentice-Hall, 1962); Daniel S. Greenberg, *The Politics of Pure Science* (New York: New American Library, 1967); Enid Curtis Bot Schoettle, "The Establishment of NASA," in S. A. Lakoff (ed.), *Knowledge and Power* (New York: Free Press, 1967).

52 See, for example, H. P. Green and A. Rosenthal, *Government of the Atom: the Integration of Powers* (New York: Atherton, 1963); H. Orlans, *Contracting for Atoms* (Washington, D. C.: Brookings, 1967); H. L. Nieburg, *In the Name of Science* (Chicago: Quadrangle, 1966); R. R. Gill, "Problems of Decision Making in Soviet Science Policy," *Minerva*, 5, No. 2 (1967); L. H. Tribe, "Legal Frameworks for the Assessment and Control of Technology," *Minerva*, 9, No. 2 (1971).

53 See, for example, R. Gilpin, *American Scientists and Nuclear Weapons Policy* (Princeton: Princeton University Press, 1962); R. Gilpin and C. Wright (eds.), *Scientists and National Policy-Making* (New York: Columbia University Press, 1963); Carl William Fischer, "Scientists and Statesmen: A Profile of the President's Science Advisory Committee," in *Knowledge and Power;* L. E. Auerbach, "Scientists in the New Deal," *Minerva*, 3 (1965), No. 4; R. M. MacLeod and E. K. Andrews, "The Committee of Civil Research: Scientific Advice for Economic Development 1925–30," *Minerva*, 7 (1969), No. 4; A. Leiserson, "Scientists and the Policy Process," *American Political Science Review*, 59, No. 2 (June 1965), 408–416.

54 Exceptions worthy of note include Don K. Price, *The Scientific Estate* (Cambridge: Harvard University Press, 1965); Alice Kimball Smith, *A Peril and a Hope* (Chicago: University of Chicago Press, 1965); J. Haberer, *Politics and the Community of Science* (Princeton: Van Nostrand, 1969); Loren A. Graham, *The Soviet Academy of Science and the Communist Party* (Princeton: Princeton University Press, 1967); H. Sapolsky, "Science, Voters and the Fluoridation Controversy," *Science*, vol. 162 (1968); 427–433.

55 David Joravsky, *Soviet Marxism and Natural Science 1917–1933* (New York: Columbia University Press, 1961).

by the remarkable growth of a journalistic offshoot. In particular, Daniel S. Greenberg has explored and mapped the frontier between science and American politics in astounding detail, and in a fashion not emulated elsewhere.[56] In focusing upon specific issues, Greenberg has shown just how dirty—that is, how like the politics of other policy-making arenas—may be the politics of science. His work did not please political leaders of American science. Alvin Weinberg, one of this elite, referred to it as "muckraking" and, while acknowledging the accuracy of the facts, suggested that it gave an inadequate picture of the relations between science and politics.[57] Though Dr. Weinberg's reaction may have been in part annoyance, in part concern that such analysis might provide ammunition for politicians ill disposed to science (as he admits), there is some truth in what he says. The picture which the "politics of science" offers is a curious one, lacking cohesion. The scientists whose names appear with monotonous regularity in the indices to such works seem to have forsaken the laboratory for the corridors of power. These men—from the Aigrains, Cherwells, and Davids to the Wiesners and Zuckermans—seem to move entirely in the company of politicians, with no contact with professional colleagues. Yet we know this is not always the case. As I hope to demonstrate, their relationships with the politicians may often be a function of their relationships with scientific colleagues—and vice versa.

Thus, it seems to me that, whereas the sociologist has usually taken an overrestricted view of science (limiting his attention to a mere section of the basic research community), the political scientist has taken an overwide one (usually taking science and technology as virtually synonymous). Whereas the sociologist has emphasized the *centripetal* forces operating in science, the political scientist has emphasized the *centrifugal* ones—that is to say, where the former has tended to exaggerate the communal nature of science (to the exclusion of external relations and effects), the latter has tended to exclude all notion (or feeling) of community from his analysis. In this book I shall take a few tentative steps toward formulation of a perspective drawing upon both these traditions. On the one hand, political scientists' findings shed a good deal of light on *ways* in which the social structure of science is dependent upon the political system in a given country. On the other hand, no longer can the sociologist neglect the external relations of the social system of science: as we move into the 1970s, so restricted an analysis is in danger of depicting science ever more inaccurately.

56 See his articles in *Science* in the period 1963–1971, his subsequent *Science and Government Report,* and *The Politics of Pure Science.*

57 Alvin M. Weinberg, "Scientific Choice and the Scientific Muckrakers," *Minerva,* 7, Nos. 1–2 (1968–1969), p. 52–63.

The State of the Art

Why should the picture become "ever more inaccurate"? The inter-actional approach to the sociology of science assumes the basic insulation of scientific institutions from those of the environing community. Is this becoming less true? Let me state the question somewhat differently. Such insulation would imply both operating freedom at the individual's level (for example, choice of research projects), as well as general autonomy of the scientific community as a whole. That is, there would have to be a level of research funding consonant with scientists' requirements, a lack of outside interference with utilization of these funds, autonomous control over access to the scientific profession, and so on. In addition, scientists would expect certain political freedoms: to discuss widely, to travel freely, and so on. Such desiderata may be found only in a social, political, and economic climate favorable to science and scientists, and in one with an insatiable demand for the results and products of research. Even then, over-riding concern over national security may take precedence, and this "high demand" situation must be regarded as a necessary but not sufficient condition.

Ben-David suggested that, although the general value system of society largely determines development of a specific scientific role within that society, one may explain differences in subsequent rates of scientific growth between societies in terms of more specific institutional factors.[58] Recipro-cally, one may explain restrictions upon scientific growth in terms of un-favorable constellations of institutional factors: rigidity of the higher education system, slight industrial and governmental demand for the results of research, central control over (and lack of competition between) re-search organizations, and so on. In the United States, the United Kingdom, and elsewhere, World War II led to an enormous demand-induced expan-sion in the scale of scientific activity, financed largely by national govern-ments. Governments acquired a new financial responsibility for science and, for the first time, scientists found themselves well supported by (but de-pendent on) politicians and bureaucrats. In the United States, for the first time, both sides had become conscious of their need for the other; in Britain a similar view, emerging under the pressures of an earlier war (that of 1914–1918), received new stimulus. Popular awareness of scientists' contribution to the war effort led to an upsurge in the profession's general status. The end of the war and the demobilization of scientists was not, therefore, accompanied by a wholesale abandonment of scientists by gov-ernment. In the United States, in particular, financial support for science

58   Ben-David, *Scientist's Role in Society,* chap. 9.

never waned. As Daniel Greenberg put it: "Perhaps the most striking feature of the first post-war decade was that the military services, recalling their pre-war neglect of science and technology . . . were now eager for an intimate relationship with the men and works of science and technology." [59] Greenberg goes on to quote James B. Conant (who had been one of the leaders of the military research effort), writing in 1952, "The Defense Department, in regard to research, is not unlike the man who sprang on to his horse and rode off madly in all directions."

Though the military departments were especially concerned with harnessing research to production of new military hardware, they were well prepared to accept the promise of long-term payoff and to fund basic research lavishly. The generosity of the federal government was no less toward basic research than toward research of more apparently short-term practical benefit.

Table 1–1

Gross Expenditure on Research and Development (GERD) in the USA

($ million)

1930	1940	1945	1950	1955	1960	1965
160	340	1,520	2,800	6,300	13,700	20,500

SOURCE: *Reviews of National Science Policy: United States* (Paris: OECD, 1968), Table 1.

Moreover, the influential Bush Report of 1945 (*Science, The Endless Frontier*) recommended a substantial delegation of authority over utilization of federal research funds to the scientific community. An agency, the National Science Foundation, was set up to finance basic research projects proposed by and evaluated by members of the scientific community. Its budget grew very rapidly, from $225,000 in 1951 (its first year of operation), to $16 million only five years later, and to $152 million in 1960.[60] However, large though the foundation's budget became, it never financed more than 10 percent of all federally supported basic research: scientists could turn equally to other agencies, such as HEW, NASA, DoD, AEC, and so on.[61] Here, from the scientist's point of view, was the strength of the

59   Daniel S. Greenberg, *The Politics of American Science* (London: Penguin Books), p. 167.

60   *The National Science Foundation: A General Review of Its First 15 Years* (Report to the Science and Astronautics Committee, House of Representatives) (1966), p. 32.

61   *Ibid.*, p. 53.

pluralist system: it offered him an abundance of sources of funds, ensuring his independence of any one political master. Moreover, as Ben-David pointed out, the federal government scarcely interfered at all: "It was learned that the best way to utilize science for non-scientific purposes was not through subjecting research or teaching to non-scientific criteria, but to aid it in its own immanent course and then to see what uses could be made of the results. . . . The link between science on the one hand and industry and government on the other was not established by the industrialists or the civil servants giving instructions to scientists. Rather there has been a constant and subtle give and take." [62] Similar generosity marked other governments' policies toward research, though nowhere (saving the USSR) on a similar scale. In Britain government expenditure on civil research (R & D) alone grew from £6.6 million in 1945 to £75.1 million in 1960.[63]

Table 1–2
Expenditure on Fundamental Research as % GERD

	1953	1963	1964	1965	1966	1967
France	—	16.5	—	—	—	18.4
Germany	—	22.5	—	—	—	23.5
Japan	—	—	—	30.1	—	28.0
U.K.	—	—	10.7	—	—	11.3
U.S.A.	9.0	—	—	—	14.0	

SOURCE: Data from OECD document DAS/SPR 70.1.

The situation changed little through the 1950s, marked principally by constant expansion in the scale of support, or through the early sixties. National governments continued to be the major source of finance for this burgeoning research effort. Throughout the 1960s, the U.S. federal government was responsible for 61 to 65 percent of total national R & D expenditure: in the United Kingdom, government paid for 50 to 57 percent; in Canada, for 50 to 60 percent; in France, for 64 to 65 percent; in Japan, for 35.5 percent (in 1959), rising to 42.0 percent (in 1967).[64] Within this rising budget, the percentage of funds devoted to basic research rose (see Table 1–2). In the United Kingdom annual budgets of the Research Councils (whose responsibilities for basic research are somewhat analogous to

62 Ben-David, *Scientist's Role in Society,* pp. 165–166.

63 Council for Scientific Policy, *Report on Science Policy* (London: HMSO, 1966), p. 20.

64 Data from OECD document DAS/SPR 70.31, p. 42 (Paris, 1970).

those of the American NSF) grew at a rate of 13 percent per annum in real terms during the early sixties. Much money provided for performance of research was used in training new scientists, and the rate of growth in Ph.D.s produced generally outstripped the growth of higher education in many countries.[65]

Table 1–3
Growth in Science and Technology Doctorates Awarded
(Putting 1960 = 100)

	1950–1	1960–1	1965–6
France	18	100	280
Japan	—	('61–2) 100	('66–7) 430
U.K.	57	100	190
U.S.A.	68	100	('64–5) 170

SOURCE: *Development of Higher Education 1950–67* (Paris: OECD, 1970). Germany has no degree equivalent to the Ph.D. and, therefore, figures cannot be quoted.

The "sellers' market"; the widespread generosity toward science; the growing emphasis upon basic research; the rapid expansion of agencies such as NSF, the British Research Councils, the German Forschungsgemein-schaft, which responded to scientists' proposals and evaluated them by means of committees of scientific peers: all these factors protected the scientist's autonomy.

Toward the end of the 1960s, as is well known, the bubble burst. Growth rates began to fall: Table 1–4 shows that expenditure rose much more slowly between 1967 and 1970 than between 1960 and 1967. The newly trained professional scientists, having started their doctoral training with very different prospects, continued to pour out of the universities— American, British, and Canadian, in particular—only to find that suitable jobs were no longer available. The "statesmen of science," in close touch with government thinking, found that they were unable to influence this new official attitude toward support of research. Then they began to urge their colleagues to accept the new situation—of general economic retrench-ment combined with popular and political disillusionment toward science —and make the best of it. Time and again, the scientific community was urged that it could no longer expect its bills to be met as of right. Don K. Price, in a 1968 presidential address to the American Association for the Advancement of Science, added the insights of a political scientist to those

---

65 S. S. Blume, *Postgraduate Education: Structures and Policies* (Paris: OECD, 1972).

of the scientists. He argued that if the scientific community was to resist the multipronged attack—the desire of politicians "to cut down on the appropriations for research . . . and to break down the degree of autonomy which the leaders of the scientific community gained a generation ago in the procedures by which research grants are distributed"; all this com-

Table 1–4
Growth in R & D Expenditure (GERD) 1961-70

	1961 ($m)	1967 ($m)	1970 ($m)	Average Annual Growth Rates (at Compound Interest)	
				1961–7	1967–70
Canada	—	828	959	14.9	5.0
France	1,004	2,522	2,765	16.6	7.4
Germany	906	2,354	3,395	17.3	10.0
Japan	678	1,675	3,395	17.3	10.0
U.K.	1,847	2,267	2,536	16.3	14.4
U.S.A.	14,552	23,680	27,250	8.5	4.8

SOURCE: Data from OECD document DAS/SPR 70.31.

bined with the ideological revolt against the scientific world view promulgated by Marcuse and others—it would have to devise radically new political strategies.[66]

William D. Carey, a senior budgetary official till 1969, has described the breakdown in relations between the science advisor and his staff in the White House (the Office of Science and Technology) and the Bureau of the Budget (now the Office of Management and Budget).[67] This was prelude to the science advisor's (then Lee DuBridge) exclusion from preparation of President Nixon's 1971 budget, a low point in the political representation of science. Mr. Carey went on to suggest that the administration hoped not only to cut back in the overall expenditure on basic research, but *itself* to impose criteria of relevance in a way it had never done before. This would affect policy toward the National Science Foundation, in particular:

Instead of the NSF being regarded as a bottomless pit into which one dropped infinite sums of money without expecting even so much as a

66 Don K. Price, "Purists and Politicians," *Science,* 163 (1969), 25.

67 William D. Carey, quoted in "Sweetness and Light Break Out Again," *Nature,* 230 (1971), 8.

splash, the NSF today is being led closer towards the market-place. When I dealt with the NSF there was a revulsion to tainting the NSF with any applied research . . . the present Administration has come to the decision that research ought to be guided towards pay-offs.[68]

And this is borne out by the subsequent preferential growth of the foundation's applied research programs.

In the United Kingdom, too, government policy toward research is now marked by a concern with economy, on the one hand, applicability on the other. The rate of growth in the science budget, around 13 percent (in real terms) in the early sixties and still over 12 percent in 1966–1967, fell to 10.8 percent in 1967–1968, and to only 4.3 percent in 1971–1972.[69] The effect of this, coupled with a reduction in funds available for support of doctoral study, has been to reduce the overall resources available to the individual scientist. The proposals of Lord Rothschild (Fellow of the Royal Society and ex-research director of Shell), recently debated so heatedly by the scientific community, have now largely been accepted by the government. I have referred to Rothschild's recommendations, formulated in the interests of both economy and applicability and likely to reduce significantly the research funds over whose allocation the scientific community holds delegated power. Lord Rothschild criticizes the extent to which decisions are made by committees of scientists within the research council structures: he suggests a transfer of power over decisions from the scientific committees to the government department with greatest involvement in an area of research. It is interesting to note the change in both general philosophy and recommendations between this report and the previous official review of the organization of British research, carried out in 1962–1963. A committee, sitting under the chairmanship of one of the country's most senior government officials, and at a time of rapidly rising government expediture on R & D, expressed its satisfaction with the extent of delegation of financial control to the scientific community:

> The Haldane Report established the principle that the control of research should be separated from the executive function of Government. The Research Council system, as we know it today, is based on this principle which, in our view, has contributed significantly to the Councils' ability to promote research and development whilst simultaneously guaranteeing the independence of the scientific judgments involved. We endorse this

68   *Ibid.*

69   Appendix A to the report *The Future of the Research Councils* (a report of a working group of the Council for Scientific Policy) in *A Framework for Government Research and Development.*

concept that public funds provided for the support of civil scientific research should be administered by autonomous Research Councils.[70]

But today the situation in Britain, as in the United States, is very different.

POLITICS AND THE SCIENTIFIC ENTERPRISE

There is considerable evidence, both theoretical and practical, for interpenetration of the political and scientific systems in modern society, and for their influence one upon the other. Yet there has been remarkably little study of this interpenetration and of this influence. Political scientists have been interested in effects of scientific knowledge (its production, existence, utilization) upon politics—but especially upon processes of executive government. The widely held view that effective scientific advance depends upon a high degree of autonomy, hitherto characterizing government policies toward basic research, has led political scientists to investigate ways in which governmental procedures have adapted to this requirement. Their emphasis has been upon implications for government of scientific development and the need of, and for, science. Questions of policy making and administration (whether explicitly for science or involving science as a means of achieving some other goal) have taken precedence over wider political issues, in terms of which science and technology have seemed much less relevant.[71] Because of this emphasis upon perturbations produced in the policy-making and administrative systems, science (more accurately, R & D) appears in political scientists' writings not as a communal activity, not as a systematic body of knowledge, but as a jumble of factors and actions. The fields of scientific research do not correspond in a neat one:one fashion with areas of governmental action.

A number of writers have drawn attention to effects of political and administrative behavior upon the scientific community. Many sociologists would follow Joseph Ben-David in ascribing postwar American preeminence in research to its pluralist system of government and to such essentially political factors as the competitive nature of its universities, the war-stimulated demand for science, the subsequent demands of the cold war, and so on.[72] The theoretical concept of "democracy," congruent with early

70 Report of a *Committee of Enquiry into the Organization of Civil Science* (Trend Report) (London: HMSO, October 1963).

71 This constraint does not typify the work of some political scientists, such as Sapolsky. See, for example, his "Science, Voters, and the Fluoridation Controversy," *Science*, 162 (1968), 427–433.

72 Ben-David, *Fundamental Research and the Universities*.

formulations of the norms of science, has been taken to characterize that
political system (the American one) which seemed most convivial to the
scientific enterprise.[73] Consequently, and for both theoretical and political-
ideological reasons, interest has attached to studying the scientific community's reactions to political environments diverging substantially from
this ideal. Robert K. Merton's essay "Science and the Social Order," [74]
which discusses effects of Nazism upon German science, is a classic and in
the 1950s, in particular, the effects of Stalinism upon Soviet scientists
attracted a good deal of attention.[75] At the same time, it became apparent
in the United States that political circumstances at home had posed, and
were posing, severe problems for the scientific community. Individuals
engaged upon military research during the war had been not merely incon-
venienced by the security barriers to their freedom, but profoundly shocked
by the decision to deploy that most powerful of their achievements, the
A-bomb, against Japan. The shock had unleashed tides of unrest within
the scientific community.[76] More recently still, scientists, like other intellectuals, were denied not only the specific freedoms deemed essential for
science, but those more elementary freedoms appropriate to life in a
democracy, as a result of McCarthyite excesses.[77]

The trial of Robert Oppenheimer served to both illuminate the tribu-
lations of the scientific community and to crystallize discontent in the field.[78]
Therefore, while there has been an implicit recognition of the susceptibility
of science, and the scientific system to political interference, this is widely
regarded by sociologists as exceptional. As I have indicated, the tendency
is to regard the social system of science in modern non-Communist societies, especially in the United States, as essentially autonomous. In particular, this implies its independence of considerations of both politics and
policy. Hagstrom, for example, on the very first page of his deservedly
influential study of the scientific community, dismisses scientists' political
activities as not relevant for the sociologist of science.[79] If I subscribed to
this view, I should never have written this book. In the twentieth century

73 Bernard Barber, *Science and the Social Order* (New York: Free Press, 1962),
especially chap. 3.

74 Reprinted in his *Social Theory and Social Structure*, pp. 537–550.

75 *Science and Freedom* (Proceedings of Conference of Congress for Cultural Freedom, Hamburg, 1953) (London: Secker & Warburg, 1955).

76 Alice Kimball Smith, *A Peril and a Hope* (Chicago: Chicago University Press,
1965).

77 Edward Shils, *The Torment of Secrecy* (London: Heinemann, 1956).

78 There is a substantial literature on the trial of Robert Oppenheimer. See, in
particular, J. Haberer, *Politics and the Community of Science.*

79 Hagstrom, *Scientific Community*, p. 1.

there may have been periods, brief periods, when scientists *did* have the best of all possible worlds: maximum facilities combined with minimum interference. Whether or not this has ever been the case, I submit that it is not so today, and I have drawn attention to aspects of government policies which reduce the validity of any assumption of autonomy. Clearly, the scientific communities of the United States and of other countries are reacting both to specific changes in government policies and to general political change in society. Among the first group of stimuli to which scientific organization is reacting are government actions in the fields of economic and foreign policy, as well as toward pure science itself. Among the second group of stimuli are issues such as the political, economic, and educational rights of minorities in society, and general public disillusionment with science both as intellectual system and as harbinger of improved welfare. While we cannot say that society is becoming less "democratic"—that is, less convivial to science—we can say that scientists seem more aware of divergences from the democratic ideal.

There are numerous indications of the scientific community's reaction to these social and political circumstances. Groups of physicists try to persuade purely scientific societies, such as the American Physical Society, to take up a public position on a purely political issue such as the war in Southeast Asia. AAAS meetings are disrupted by scientists who resent the discrepancy between its constitutional commitment to the promotion of human welfare and its practical inaction. For the first time since the office was instituted by President Eisenhower, a president's scientific advisor (Lee DuBridge) finds it impossible to function as the servant of two masters: answerable simultaneously to scientific community and political superiors. Subsequently, the office itself was abolished. But these much-publicized incidents are no more than the visible tip of the iceberg, serving to attract our attention to the hidden bulk below. The presumption that, in terms of structure and interactions, the scientific system may be regarded as insulated from external influence has led sociologists to ignore many of the most interesting aspects of scientific behavior and organization. Consider development of a society formed in response to the needs of a newly emergent science. Sociologists have drawn attention to the strains which such a society may feel as a result of differentiation within the discipline. Organic chemists, for example, may believe that a society covering the whole of chemistry, with its associated journal, cannot do justice to their specific demands. They may seek to form their own group, with a journal devoted solely to organic chemistry. The original society may seek to accommodate their demands to avoid their total secession.[80] But history shows that these strains, and the responses to which they gave rise, were almost exactly matched by the

80 Hagstrom, *Scientific Community,* especially pp. 187–226.

effects of growing professionalism. In other words, as I shall show later, economic changes may have similar implications for scientific organization as changes in the internal structure of the field which have monopolized the attention of sociologists of science.

In the following chapters I shall attempt to sketch out some sociological issues evidenced by discarding the assumption of the autonomy of science and focusing attention upon relations between scientific and political systems: upon sociopolitical dependence of the scientific role. Chapter 2, the most explicitly theoretical, is concerned principally with the conceptual nature and institutional norms of science. The latter, in Merton's words "moral as well as technical prescriptions," may be seen as not only functional for achievement of the institutional goal of science, but as having a certain psychological appeal for the committed scientist. Two philosophies of science, offering rather different notions of scientific progress, are discussed. The use of Kuhn's conceptual scheme is shown to be but partly compatible with Merton's formulation of the norms of science. Merton and Barber's identification of the values of science with those of democratic society is discussed and compared with the very different values imputed to science by Marcuse and Habermas. Variations, it is suggested, result both from very different conceptions of what science "really is" (philosophies of science) and from very different perspectives on American society, which has proved so hospitable to science. But, whichever view one accepts, science must be seen as innately political. It can also be shown that from the practical interrelations of science and politics follow certain political roles for scientists.

In Chapter 3 I consider the operation of the scientific system. How is social control exercised within the scientific community, and by whom? The commodity of the scientific exchange system, recognition, is discussed in its various forms, and their allocation considered. Scientific norms prescribe allocations of rewards solely in proportion to contributions made to science, but there is evidence that some individuals profit from entirely extraneous statuses. Some of these factors, such as academic background and affiliation and political views, are discussed. Finally, I consider briefly the situation of scientists in environments in which organizational control is imposed, wholly or partly, in place of professional control. Organizational scientists' responses to a situation in which they feel little freedom in choice of research (for example) may be either accommodation (as a number of sociologists have pointed out) or, more aggressively, a militant attempt to secure a measure of self-determination.

Chapter 4 is concerned with the kinds of organizations formed by scientists. The function, genesis, and evolution of scientific societies is illustrated with reference to chemical organization in the United States and in the United Kingdom. Subsequently, both American and British chemical societies were faced with crises: one deriving from the process of disciplinary differentiation (as with the growing discontent of organic chemists), the other from emergent professionalism, as chemists were newly absorbed by the economic system. Each crisis called for some kind of accommodative reaction. And later—notably in the 1920s and again today—the professional scientific societies would face new problems, deriving from economic, political, and social change.

In Chapter 5, I discuss the notion that science today is in crisis, and some currently observable effects. I suggest that the nature of scientists' discontent has led professional men to sacrifice their professional status (however reluctantly) for a more militant trade union orientation. Typically, "unionization" consists, on the one hand, of the increasing militancy of preexisting professional organizations and, on the other, of the formation of new, more radical bodies. Moreover, two orientations in the nature of union-type organizations may be distinguished. The first "instrumental unionism" emphasizes such practical desiderata as bargaining rights, higher salaries, and so on. The second, "ideological unionism," emphasizes the need for radical change in the social and economic structure of society. A clear tendency among scientists in the United Kingdom may illustrate the first kind of unionism, while in the United States the growth of such bodies as the Federation of American Scientists and Scientists and Engineers for Social and Political Action may illustrate ideological unionism. These and other radical science organizations are discussed in some detail.

The government scientific advisory system provides the major focus for Chapter 6. We discuss the needs of governments for authoritative scientific advice and the functions of advisors. How, and based on what criteria, are individuals selected to perform these functions, so prestigious in the scientific community? In what ways is the resultant group "representative" of this community which, in a sense, it represents in the councils of government? If government contacts with the scientific community are largely restricted to an elite, what implications does this hold for relations of the elite with the community?

A comprehensive treatment of the political relations of the scientific system (which, I hasten to add, I do not profess to have produced) could not exclude from consideration relationships obtaining outside the executive arm of government. Both chapters 7 and 8 consider links between the scientific community and the community-at-large. The commitment of the scientific community to increasing public understanding and awareness of science is institutionalized in the existence and objectives of such bodies as the British and American Associations for the Advancement of Science. How, in fact, do citizens acquire their knowledge of scientific matters, and how much knowledge do they possess? The mass media provide a major channel of communication, albeit one often accused of distortion. What are characteristics of the mass media's treatment of science? Finally, one relatively contentious issue in which the public voice was heard, the question of fluoridation of water supplies, is discussed with reference to the oft-mentioned problem of popular involvement in science and technology policy determination. This also illustrates, however sketchily, another facet of the political relations of science.

All this is no more than a preliminary sketch and, in making it, I am all too aware of crucial gaps in knowledge—and, more particularly, in my own knowledge. I can but hope that the perspective I propose will commend itself to those with both greater insight and greater facility for empirical research than I can claim. The issues raised in the following pages would appear not only clearer, but in their true perspective, in the light of such enlightened inquiry.

chapter 2
Science and Political Values

In this chapter I shall be concerned mainly with consequences of the precepts to which scientists may adhere in carrying out basic research. These precepts, rules of conduct, "institutional norms," both prescribe and describe the conduct of scientists. They are prescriptive because, it has been suggested, by adherence to them individuals are best able to advance the cause of science. They are descriptive—though not necessarily of the behavior of any one scientist at any particular time—because they help us understand the group behavior of scientists. These institutional norms, moral as well as technical prescriptions, serve to orient the scientist toward the institutional goal of science: therefore, they are dependent for their meaning upon this goal. As Merton puts it, "The entire structure of technical and moral norms implements the final objective." [1] So, in order to understand them, to appreciate their importance for the scientist, we must develop a working conception of what science is all about. This is a problem with which philosophers have long grappled and, together with the subsidiary question of what gives science its unique "progressive" character, it provides the central subject matter of the philosophy of science.[2]

1 Robert K. Merton, *Social Theory and Social Structure* (New York: Free Press, 1957), p. 552.

A brief discussion on the philosophy of science may offer additional benefits. In proceeding to discuss science as a "social system," in examining or formulating a sociological approach not to scientists in the work groups in any laboratory but to those supposedly united by the bonds of "community," we are making an important assumption. We are *assuming* that it is right and proper to regard scientists in this way. Any sociological evidence we may amass will be based upon this assumption and cannot ever be said to validate it. Therefore, it becomes of interest to seek support for such an assumption from an external source, and philosophy may provide such support. What support, if any, does the philosophy of science provide for conceptualization of the field as a social activity, in which scientists are united by something other than the face-to-face interactions of work groups? In simplified terms, two major viewpoints are current within the philosophy of science: one regards scientists as involved principally with formulation and refutation of hypotheses; the second sees them as engaged mainly upon the much more mundane activity of "puzzle-solving." I propose to discuss them now, their relationships to one another, and their utility for the sociology of science.

The first view, associated with Sir Karl Popper and his school, regards science as essentially creative and critical, proceeding from theory to theory by a rational process of critical discussion, refutation (rather than validation, which is logically impossible), and reformulation. For Popper, a theory must be better than its predecessor (where "better" implies, for example, greater scope), and may then be regarded as "nearer the truth." Therefore, science approaches a *true* view of nature. The usual name for Popperian theories of science is "falsificationism": it is useful, for present purposes, to refer to them as "critical-creative." A recent and perhaps somewhat deviant version of Popperianism is due to Professor Imre Lakatos.[3] Lakatos suggests that scientists are less ready to discard a still useful theory, in the light of apparently falsifying evidence, than Popper

2   This is not to suggest that professional philosophers of science have in any sense possessed a monopoly of meaningful views of what science is about. Karl Marx had very different views, so today do philosophers such as Herbert Marcuse. Later in the chapter I shall discuss some of these contrary views. Philosophers of science alone have sought systematically to discuss these problems and, though we may find their perspective restricted or deficient, it seems an appropriate place to begin. For a general treatment of "views of science," see J. R. Ravetz, *Scientific Knowledge and Its Social Problems* (London: Oxford University Press, 1971), chap. 1.

3   The major source of Popperian theory is his great work *The Logic of Scientific Discovery* (London: Hutchinson, 1959). My interpretation of Lakatos' views is taken entirely from his "Falsification and the Methodology of Scientific Research Programmes" in Lakatos and Musgrave (eds.), *Criticism and the Growth of Knowledge* (London: Cambridge University Press, 1970), pp. 91–196.

seemed to imply. According to Lakatos, while the objective of science is the substitution of one theory t_2 for another less good one t_1, scientists neither do nor should discard t_1 in the light of apparently falsifying evidence (more strictly, we should refer to falsifying "empirical propositions") *unless* an alternative, t_2, is available. Lakatos speaks of saving theories which retain their heuristic power (power to suggest new, corroboratable facts) by invoking "auxiliary hypotheses which satisfy certain well-defined conditions" in order to harmonize them with the contradictory evidence. This he terms "methodological tolerance." The term "research programme" is used to describe a series of theories and their associated methodological prescriptions for research, and it is a healthy situation for rival research programs to coexist competitively. All this is a recent gloss on the strict Popperian orthodoxy: the key point is the emphasis placed by these philosophers upon the "critical-creative" character of science. Both Popper and Lakatos stress the "normative" character of philosophies of science, denying the relevance of sociological or psychological studies of what scientists *actually do* to their formulation.

The second view refers to that explicit in the work of Thomas Kuhn,[4] to which I have referred in Chapter 1. Kuhn argues that scientists are not usually concerned with formulation of hypotheses or theories, but with solving "puzzles." The puzzles appropriate for investigation derive from the attempt to elaborate theoretical aspects of a generally accepted "paradigm." Scientists, for Kuhn, belong to communities with unique intellectual, theoretical, and normative traditions. Under normal conditions these communities are largely isolated from outside influences, and puzzles upon which members work are determined by consensual application of the paradigm. It is this process, the solution of problem after problem within a generally agreed framework and in generally agreed ways, which for Kuhn gives science its progressive character. This, in his view, distinguishes the sciences from other intellectual activities lacking the "progressive" nature of science. Scientists are trained for this kind of problem- (or puzzle-) solving activity. Occasionally, crises arise within scientific (disciplinary) communities. Problems may appear which seem to resist solution in the light of the currently employed paradigm, casting doubt upon its continued value. The consensus may be destroyed. New kinds of hypotheses may then need formulation and testing for utility and, at such critical times, scientists may turn outside the immediate confines of their discipline in search of plausible, useful theories. Under these "revolutionary" circumstances, scientists' activities appear to correlate with Popper's description. Kuhn's view is sociological,

---

4 Especially *The Structure of Scientific Revolutions* (Chicago: Chicago University Press, 1962).

as he readily admits. His scientists are organized into communities mainly insulated from outside influence, seeking to communicate solely with one another. Scientists are vastly more community-directed than practitioners of other intellectual disciplines. In Kuhn's view, thus, Popper characterizes the scientific enterprise in terms applicable only to the occasional revolutionary periods. He argues that a process of continuous criticism, of continuous revolution, cannot produce progress—being much more akin to the arts than to the sciences.[5] Popper and Lakatos criticize Kuhn on many grounds. They do not deny that some scientists are engaged in "puzzle solving," or that "normal" science exists. "It is the activity of the non-revolutionary, or more precisely, the not-too-critical professional. . . . The 'normal' scientist, in my view, has been taught badly. I believe . . . that all teaching at the University level . . . should be training and encouragement in creative thinking. . . . The normal scientist . . . has learned a technique . . . he has become what may be called an *applied scientist,* in contradistinction to what I should call a *pure scientist"* [6] For Popper, normal science has been irrelevant to the progress of science as "recorded by the history of science." Lakatos writes: "The history of science has been and should be a history of competing research programmes (or, if you wish, 'paradigms'), but it has not been and must not become a succession of periods of normal science: the sooner competition starts, the better for progress." [7] Popperians see Kuhnian theory as dependent upon a notion of the *commitment* of scientists to a specific theory within periods of normal science. Since they regard "commitment" as unscientific,—opposed to their own crucial notion of criticism,—they reject such theories.

What are we to make of these two approaches, both as rival attempts at dealing with the question "What is science?" and as aids to our own research program? Kuhn stresses conformity, "inward-orientation" (a science largely insulated from the general intellectual currents of society), puzzle-solving. Lakatos stresses the criticism and formulation of increasingly powerful theories, denying the meaningfulness of fact-gathering and puzzle-solving. So far as the professional philosophy of science is concerned, resolution of the dispute is likely to come (if it comes at all) from the more and more detailed examination of the history of science, since both sides are agreed upon the primacy of historical evidence. So far as we sociologists of science are concerned, Kuhn's thesis has many advantages. Since it offers

5 T. S. Kuhn, "Reflections on My Critics," in *Criticism and the Growth of Knowledge,* p. 244.

6 Karl Popper, "Normal Science and Its Dangers," in *Criticism and the Growth of Knowledge,* pp. 52–53.

7 Lakatos, "Methodology of Scientific Research Programmes," p. 155.

us a philosophy of science based upon an essentially social conception (not just on what scientists do when they are not being scientific [8]), it has an obvious value for the sociology of science. Kuhn's perspective suggests that we may reasonably regard science as an autonomous social institution. A view of science emphasizing its acritical, rational, puzzle-solving nature is probably most in accord with the popular view of science (certainly distinguishing it from the arts) and thus has potential value in discussing science as an ideology (as I shall do below). Kuhn argues that scientists are trained for normal science—trained, that is, to solve puzzles, to engage in a "highly convergent activity." [9] Not only is this proposition empirically testable, but it suggests a continuity between the activities of pure scientists (with whom alone philosophers of science are concerned) and those of applied scientists (who cannot be dismissed out of hand by the sociologist). There is no reason not to suppose that scientists and technologists share common training procedures as well as common social structures and technical (if not moral) norms. It would be impossible to find a link to applied science in the Popperian corpus. So, although Kuhn seems less certain of the ultimate goal of science, concentrating rather upon its "marginal" progress, let us accept his view for the moment as a background for understanding the norms of science.

The "logical status" of the norms of science, like that of the tenets of the philosophy of science, is (as I have already argued) both descriptive and prescriptive. Thus we cannot accept the Popperian's implied criticism of a sociology of science:—that it does no more than describe the day-by-day activities of a mass of mediocre fact-gatherers. Therefore, there should

---

8 Lakatos' view also has a sociological or psychological component, found in his notion of "methodological toleration." Whether or not the phenomenon in question is best described in these terms or as "commitment" is far from apparent, as the following example may help to make clear. Recent experiments suggest the arrival at the earth of pulses of gravitational radiation, apparently deriving from conversion of mass to radiant energy at the center of the galaxy. But the rate at which it seems to arrive seems incompatible with the theory of general relativity, given present estimates of the age of the galaxy. Do we then reject relativity theory? "I am not so sure. It seems to me that one should give up the accepted laws of physics only after a severe struggle has shown that they cannot account for any new observations which threaten them . . . we have had Weber's results for only two or three years, so I do not think that we should give up yet on trying to explain them in terms of general relativity." Dr. Dennis Sciama, "Cutting the Galaxy's Losses," *New Scientist* (February 17, 1972), p. 373. (This extract first appeared in *New Scientist,* the weekly review of science and technology, 128 Long Acre, London WC2, and appears with the publisher's permission.)

9 This fits in with Hudson's empirical finding that clever British schoolboys specializing in science were largely convergers (they scored better on IQ than on creativity tests), those specializing in arts were largely divergers. See Liam Hudson, *Contrary Imaginations* (London: Methuen, 1966).

be no logical objection (on the grounds of incommensurability) to formulation and interpretation of these rules for action in the light of (philosophical) discussion of the goals of science. Moreover, when, as in Kuhn's
philosophy, the progress of science is seen as an essentially corporate endeavor (as, to some extent, it is in Lakatos' also—as when he emphasizes
the need for competition), rules for *social* action would seem as inevitable
as rules for *individual* action. As I have suggested, further support for the
idea of a sociology of science is provided by Kuhn's description of what a
sociologist would regard as an autonomous subsystem of social action. In
the remainder of this chapter, I shall discuss social aspects of science, starting with those characteristics fundamental to empirical studies by sociologists (further discussed in Chapter 3), then taking up again formulation of
the norms of science by Robert K. Merton and Bernard Barber. Discussion
of the political implications of science may then be properly introduced by
Merton and Barber's attempts at demonstrating an apparent congruence
between the norms of science and the norms and values of Western "democracy."

THE SOCIAL SYSTEM OF SCIENCE [10]

The word *system* implies interrelatedness: no change in one part of a
system fails to involve implications for all other parts. Sociologists use the
term to stress *patterns* of interaction between individuals characteristic of
the system in question, recognizable, and not subject to unanticipated
change. Unlike the situation in physical systems (an engine, for example,
or an isolated "adiabatic" system of gas molecules), we must allow human
actors a degree of choice in determining to act in those ways typical of
political, economic, or family life. Talcott Parsons has concluded that
maintenance of these characteristic patterns of action is indicative of their
mutually rewarding nature for participants. The permanence of social systems and of economic, religious, and other subsystems indicates that the
rewards received by participants in response to prescribed behavior are
unchangingly desirable and obtainable. Thus, if we want money we must
work, if we want salvation we must pray, if we want political influence (or
the semblance of influence) we must support our chosen candidates: then
the free working of the system allocates us our due rewards. Some sociologists have emphasized the "exchange" element in the operation of social
systems,[11] which are then regarded as regulators of the exchange of, for
example, work for money. Norman Storer deduced three general principles

---

10 In the following section, I shall present what is essentially the structural-
functionalist view of science. There are, of course, others (as will become apparent).

11 P. M. Blau, *Exchange and Power in Social Life* (New York: Wiley, 1964).

governing the workings of any such exchange system (subsequently, he applies them to an analysis of the working of the scientific subsystem).[12]

1. The reward characteristic of the system must be universally desired within the system, and participants must be able to assume that it is desired by others.

2. The reward is to be obtained only within the system, via the due process of interchange. This is to say that many potential ways of securing money or political influence are proscribed.

3. In particular, participants must not confuse the rewards of the various social subsystems within which they function. It is not, for example, permitted to seek political influence in exchange for money, or religious salvation in exchange for political influence.

Implicit in this formulation are the need to maintain subsystems of social action and the autonomy of these subsystems. It may be characteristic of certain societies that these two imperatives come into conflict: to maintain its fundamental mission, worker-priests carry the church into the factory; government intervenes in the economy to protect the state against economic collapse; the family becomes an agent of political domination (in some psychopolitical theories). Such restrictions of autonomy may be necessary to maintain systems of interaction; on occasion, it may be impossible to understand the continuing operation of subsystems of action without recognizing that such compromise has taken place. I shall suggest that this is the case with the scientific system: hence, the notion of a "political sociology" of science. A theoretical framework of this kind, involving interpretation of social structures in terms of relations between individual subsystems, underlies much of the current sociology of science. This is especially true of Storer's major attempt at theoretical elucidation of the social system of science. His analysis depends upon the assumption, common to all Mertonian sociologists, that science may be understood as exchange of a characteristic commodity for a characteristic response, under conditions regulated by a specific set of norms. Many difficulties are involved in using an exchange model of this kind to analyze the social system of science.[13] One, which has received relatively little discussion, is the precise nature of the commodity of science: what do scientists contribute to the scientific community, for which they expect reward?

12 Norman W. Storer, *The Social System of Science* (New York: Holt, Rinehart & Winston, 1966), chap. 3.

13 R. Collins, "Competition and Social Control in Science," *Sociology of Education,* 41, No. 2 (1968), 123–140.

The Commodity of Science

Sociologists have employed relatively unsophisticated concepts of what scientists try to do, even when unconstrained by restrictions of empirical expediency. They are regarded as attempting to extend the body of certified knowledge, *tout court*. Hagstrom sees the contribution of information as crucial, although he points out that such information must be original for the higher rewards.[14] Merton emphasizes the importance of originality and the significance of establishing one's own priority in making a discovery. Because originality alone is rewarded, the history of science is replete with examples of so-called priority disputes.[15] Even today, when journals record the exact time at which a manuscript is received, disputes may rage. For example, a long-standing dispute between physicists at Berkeley (U.S.) and at Dubna (USSR) over the discovery of element 104 has recently been succeeded by one over the primacy in discovery of element 105. Such a dispute can become quite philosophical, as the nature of discovery is often not clear cut:

> I would like to raise the question of what constitutes the discovery of a new element. It seems to me that the discoverer is the one who first proves that he has indeed found a new element. Our published work demonstrates beyond question that we have identified the isotope $^{260}105$ by linking it genetically to its well-known laurentium daughter ^{256}Lr. . . . On the other hand the Dubna discovery of a 2-second spontaneous fission emitter is still open to question as to the identity of the atomic number involved.

This dispute has been conducted in the correspondence columns of *Science*.[16] Mertonian sociologists have emphasized the importance of priority: since original information is rewarded, priority disputes are necessary for maintenance of the scientific system. They can also produce a neurotic anxiety in scientists, and deviant patterns of secretive behavior. Hagstrom found that 30 to 40 percent of American scientists in some fields expressed concern over being "scooped" in their current work, to the extent that it could inhibit their willingness to enter into free discussion.[17] Similar results have

14   W. Hagstrom, *The Scientific Community*.

15   R. K. Merton, "Priorities in Scientific Discovery: A Chapter in the Sociology of Science," *American Sociological Review*, 22 (1957), 635–659.

16   Letter from G. N. Flerov, *Science*, 170 (1970), 15; Letter from A. Ghiorso, *Science*, 171 (1971), 127 (quoted above). (Copyright 1971 by the American Association for the Advancement of Science.)

17   W. O. Hagstrom, "Competition and Teamwork in Science" (unpublished mimeograph, University of Wisconsin, 1967), chap. 1.

been found in a study of British chemists.[18] Storer has adopted a somewhat different, essentially Popperian perspective.[19] In his view, it is important to recognize that science is a creative activity: scientists are, individually, creative in the same way as artists. The explanation of this creativity must be in psychological terms, but it is a phenomenon with an important social dimension:

> The creative individual works in relation to a set of standards that exist apart from the creative process and that are essentially social rather than psychological.[20]

One aspect of this is the need to produce something new to the world, not just to the self. From this follows the crucial nature of communication because, for Storer, creators need to secure validation of their creation as well as of the feeling developing with it. A significant difference between science and other creative activities is that scientific creation is aimed at a restricted audience of specialists, themselves creators, and it is from them that validation is required. The same is said by Storer, as by Kuhn, not to be true of nonscientific creation.[21] This fact ensures a *competent,* trustworthy response, also guaranteeing an exchange between equals, endowed with a *communal* dimension. Clearly, the presentation of a creative act is not the same as the presentation of information. Though these differences in ways of conceptualizing what scientists try to produce exist at the theoretical level, inadequate though they may be, they have scarcely found their way into empirical studies. Many of these studies, seeking quantitatively to investigate the exchange process, have taken "productivity" as their measure of what is given. In such studies, productivity has generally been related to the number of a scientist's publications, whether all treated as equivalent or divided into "major" and "minor" publications. This simplification is resented by scientists, as anyone will testify who has attempted an exercise "tainted" by paper counting. And yet, if "quantity of information" was really involved, this would seem an unobjectionable

---

18 S. S. Blume and Ruth Sinclair, "Competition in a Scientific Discipline: Some International and Intradisciplinary Differences" (unpublished).

19 Storer, *Social System of Science,* chap. 4.

20 *Ibid.,* p. 60.

21 This may be an inadequate distinction today. While the audience for any scientific discovery may have now become relatively small, and it may be that the discoverer is only interested in communicating with this small audience, the same phenomenon might be observable with much artistic creation. One talks increasingly of a "painter's painter," a "director's director." And in philosophy there is substantial disagreement between those who would talk only to philosophers and those who see a wider role for their work, such as Stuart Hampshire.

measure. Clearly, this is not what scientists think they are doing. Other sociologists have attempted independently to assess both quantity and quality of information.[22] The Coles found these two factors broadly correlated although, when they were not, quality was the more highly prized. But the idea of "information"—even if weighted by quality or originality— is indicative of an inadequate understanding of the nature of pure science. That sociology need not be limited by such empirical constraints (by what can be measured implying, inevitably, information) is shown by Gaston's discussion of functional differentiation in high-energy physics.[23] There are theoretical high-energy physicists and experimental ones, and they do different things. Production of new information is the prerogative of the experimentalists: they it is who run the accelerators, and whose bubble chambers produce the data-yielding film. Theorists produce no new information: they work only with the proverbial pencil and paper (and maybe a computer!). And yet, when Gaston attempted to see if there were status differences between the two kinds of physicists, he found that theorists obtained more recognition for their contributions than did experimentalists publishing the same number of papers.[24] Clearly, then, theoretical construc- struction may be as highly valued by the scientific community as new information. Interestingly, Kuhn regards the question of if and when this is the case as primarily susceptible to sociological analysis,[25] although we may feel he underestimates the contribution of the philosopher at this point. The relative importance of theory and data to the progress of a scientific discipline at any one time must be a function of its intellectual structure: it is possible to conceive of a discipline requiring above all formulation of some central theory, and of another in which "normal" scientific advance depends mainly upon collection and analysis of factual material. It is reasonable to assume that contributions are rewarded in proportion to the special needs of the discipline at the time, whether for theoretical insight, data, or methodology. Such an assumption would seem independent of one's theoretical perspective on science, depending merely upon the supposition that "scientists know best." In reacting against a view which, positivistically, has seen progress as dependent solely upon collection of falsifying (or even validating) data, Storer has gone too far in emphasizing the "divergent" or creative aspects of science and the intellectual (if not structural) parallels

22 S. Cole and J. R. Cole, "Scientific Output and Recognition: A Study in the Operation of the Reward System in Science," *American Sociological Review,* 32 (1967), 377–390.

23 J. C. Gaston, "Big Science in Britain: A Sociological Study of a High Energy Physics Community" (Ph.D. dissertation, Yale University, 1969).

24 *Ibid.,* pp. 145–146.

25 T. S. Kuhn, "Logic of Discovery or Psychology of Research?" in *Criticism and Growth of Knowledge,* p. 21.

between science and the arts. In spite of the Popperians' normative strictures, at least a measure of credit accrues to the filler-in. A true sociological understanding of the structure of a scientific discipline requires appreciation of both its intellectual structure and of functional differentiation, whether between theoreticians and experimentalists or on some other basis.

Communication in Science

A good case could probably be made for seeing any exchange process in terms of "communication": one may note Karl Deutsch's study of the political system,[26] and the rather more extreme formulations of Marshall McLuhan.[27] In science the case would be exceptionally strong, for "information" (whether in the form of experimental observation or theory) is the basis for the exchange. The theoretical physicist John Ziman has made this point with particular clarity: "An investigation is by no means completed when the last pointer reading has been noted down, the last computation printed out and agreement between theory and experiment confirmed to the umpteenth decimal place. The form in which it is presented to the scientific community, the 'paper' in which it is first reported, the subsequent criticisms and citations from other authors and the eventual place that it occupies in the minds of a subsequent generation—these are all quite as much part of its life as the germ of the idea from which it originated. . . ." [28] Another who has emphasized the importance of communication in science is Derek Price.[29] Price has devised an historical study of science, the "science of science," founded upon the quantitative study of scientific communication. His exponential curves, though giving rise to incalculable misunderstandings, have demonstrated the contemporaneity of science. In the absence of an independent theoretical framework, however, it could be argued that Price attaches too much weight to the significance of communication.

Studies by Menzel,[30] by Garvey and Griffith,[31] and by others have also demonstrated the importance of nonwritten means of communication between scientists. The pace of modern research and the growing delays in

26 K. W. Deutsch, *The Nerves of Government* (New York: Free Press, 1963).

27 See, for example, his *Understanding Media* (London: Routledge, 1964).

28 J. M. Ziman, *Public Knowledge* (London: Cambridge University Press, 1968), p. 103. © 1968 Cambridge University Press.

29 D. J. de S. Price, *Science Since Babylon* (New Haven: Yale University Press, 1961); *Little Science Big Science* (New York: Columbia University Press, 1963).

30 H. Menzel, *The Flow of Information Among Scientists* (New York: Columbia University Bureau of Applied Social Research, 1958).

31 W. D. Garvey and B. C. Griffith, "Scientific Information Exchange in Psychology," *Science*, 146 (1964), 1655.

publication have bestowed increasing importance on symposia, travel, circulated preprints (mimeographed versions of papers as yet unpublished), and so on as sources of information. The continued survival of the journal throws into relief its archival function in modern science—not to inform, but to record achievement. The informational value of personal contacts and so on is demonstrated by the anguish and disaffection of many scientists,—in developing countries isolated by lack of travel facilities—unable to attend international jamborees [32] and thus make contact with their peers. As a result, much research carried out in developing countries is no more than a duplication of work done elsewhere. The continuing importance of journals, in spite of forty years of incipient mourning, is demonstrated by the fate of attempts at premature burial. One such, exceedingly controversial, has been the attempt by the American Psychological Association (APA) to reform the communications system of that discipline. In place either of journals, or of preprints distributed personally by scientists, the APA scheme involves circulation by a central clearing house of unedited or refereed papers to all signifying their interest in the requisite area of research.[33] This "early warning" system is designed to overcome problems arising from proliferation and growth of primary journals requiring scanning by scientists, delays in publication, high costs of journal production, and specific needs of the association's nonresearch members. On the new system, manuscripts would be distributed as received within two months of receipt. Frequently, critics of the proposals make the point that not only would all this result in a greater volume of circulating documentation but that, in addition, it would increase the concentration within this volume of sheer rubbish.

A different kind of criticism is voiced by James J. Jenkins (Professor of Psychology at the University of Minnesota), who resents the transfer of responsibility for the communication system of psychology from practitioners to engineers and computer documentation specialists. The view that "scientific communication is too important to be left to the scientists" finds little favor. This debate—and there have been other similar ones—is indicative of two things. In the first place, scientific communication is rather more than circulation of data—rather more, indeed, than circulation of the results of science. Second, following from this, control over organization of communication within a discipline is a basic aspect of scientific autonomy. Control over the journals, and over the informal communication system, may be entirely different requirements. By considering the problems of emergent

---

32   A. Salam, "The Isolation of Scientists in Developing Countries," *Minerva*, 4, No. 4 (1966), 461–465.

33   "Psychology: Apprehension over a New Communication System," *Science*, 167 (1970), 1228.

specialisms, Hagstrom illustrated the need to publish, and the use of journal control as a form of sanction over those working in deviant areas.[34] So strong is the need that restrictions on publishing opportunities can lead to formation of a new journal, whose existence encourages identification with the new field and is a prerequisite for such identification. And yet, almost certainly, those working in this deviant specialism will have created their own informal communication channels, linking them to each other. They will have constituted their own invisible college and, if restricted access to media of publication is a "push" stimulus to formation of a new journal, operation of the invisible college may frequently act as a "pull" stimulus. Growth of the informal group or "network," [35] an increasing need to be selective in terms of both men and documents, these are the signs of change and, as Garvey and Griffith have pointed out, they may be signs of more in addition: [36]

> This process may continue to evolve until someone realizes that an institution with most of the characteristics of an archival journal has emerged— a large and increasing input of manuscripts, an existing gatekeeping group, an eager and expanding audience, and growing economic problems. And thus a new journal is born!

So the communication system of science is much more than it appears to information experts. Its structure reflects and supports the structure of science itself, and protection of the autonomy of science involves protection of this complex structure. Within all this, in which each element has its own part, the published paper assumes a special significance for the scientist. A published paper is an acknowledgment that he has made a worthwhile contribution—but it is the hopeful young aspirant to whom we should turn for a real understanding of the significance of publication. We should ask him about his feelings on publishing his first paper.

The Reward System of Science

Discussion of a social system in terms of an exchange process must attach central importance both to what the individual contributes to the system and to what he receives in return. The empirical part of such discussion can reasonably treat the working of such exchange, the autonomy of this reward system, for example, and the extent to which allocation of

---

34 W. Hagstrom, *Scientific Community*, pp. 108–110.

35 N. C. Mullins, "The Development of a Scientific Specialty," *Minerva*, 10, No. 1 (1972), 51–82.

36 W. D. Garvey and B. C. Griffith, "Communication in a Science: The System and Its Modification," in *Communication in Science* (London: Churchill, 1967).

rewards correlates with the norms of the system. Although from all this derive a considerable number of fundamental questions, hitherto the sociology of science has addressed itself to but a limited number. Just as there has been only the most rudimentary discussion of the content of scientific communications, as distinct from their function, too little attention has been paid to the real nature of the rewards of science. Merton, and following him most others, has considered that the reward system of science is founded upon giving "recognition and esteem to those who have . . . made genuinely original contributions to the common stock of knowledge." [37] According to Merton, therefore, what scientists want is a public assessment and acknowledgment of the value of their contribution. Norman Storer's view, as indicated above, is somewhat different. He feels that what scientists seek in making public their work is not professional assessment of their contribution but, rather, a "competent response." The creator needs legitimation of his feelings of creation, and of what he has created, by individuals whose competence he can trust. The nature of the evaluation of his work, the extent to which it is deemed worthy of reward, is of less importance.

Although scientists may not conduct their researches with the primary aim of securing recognition, but must feel bent upon contributing to the body of knowledge, the exchange theory of science presupposes that the reward which the scientific community can offer in return is wanted. If scientists had little regard for acknowledgment by peers, would they not work exclusively in private industry for some enterprise which would reward in money (they may be presumed to value this) their contributions to profitability? Some do just that, and their situation has been a matter of dispute among sociologists of science, as we shall see below. But discussion of the ideal system of science has not centered on this. Almost all empirical research on the reward system of science has been founded upon the assumption that it is directed toward allocation of professional recognition. There has been no attempt, so far as I am aware, to test Storer's, or *any* alternative, view. It is, then, proper to ask why the system works:—motivational as well as institutional questions.

Why do scientists so value assessment, and appropriate recognition by their peers, of their work? One answer may be in terms of socialization into the scientific role, involving growing acceptance of both the necessary intellectual commitment and its social imperatives.[38] In studying science, in learning to become scientists, students learn to desire the rewards of the scientific life from their committed masters. Storer is not satisfied with an explanation of this kind. He must ask, "Why should anyone choose to enter upon this arduous training process?" It is necessary to ask what traits of

37   Merton, "Priorities in Scientific Discovery."
38   Hagstrom, *Scientific Community*, pp. 9–12.

certain individuals lead them to enter upon the socialization/training/educational process which after many arduous years turns them into scientists.[39] In Storer's view, a comprehensive sociology of science "must take account of the *psychological* foundations of the reward system of science." An explanation resting ultimately upon socialization is also rejected by Cotgrove and Box, but for rather different reasons. Statistics show that substantial percentages of those completing each stage of a scientific training (from undergraduate through to doctoral and even, to a lesser extent, postdoctoral work) demonstrate, by their choice of occupation, a lack of commitment to basic science. Why is it, Cotgrove and Box ask, that some students feel committed to science by the time they leave the university, whereas others do not? [40] The answer, they suggest, lies in a connection between occupational choice and socialization: students' perception of future career influences the socialization process deriving from and accompanying higher education. Therefore, Cotgrove and Box are driven to investigate a wide variety of external factors which might predispose toward the scientific career and thus explain the socialization process. Part of the explanation is in terms of marginality. The experience of higher education in the essentially middle-class British universities may constitute a marginal situation for working-class students. These authors found that British working-class students were more committed to the scientific career than were middle-class ones partly, they suggest, because of the classless image of the scientist, partly because of the emphasis upon technical rather than social skills.[41] The question of *why* scientists want professional recognition, and why they are prepared to "stick to the rules" to obtain it, is difficult and not directly relevant to the argument of this chapter. Although there is some disagreement over whether the fact that many seem to have "sacrificed" their scientific integrity by taking industrial employment should be regarded as a natural or a strained situation, there is no doubt that, generally, much of the basic research community works in the prescribed fashion. The behavior of both individuals in carrying out and publishing their work, and of the scientific community in evaluating and rewarding the individual contributions, approximates that regarded as "proper."

The Norms of Science

Now I want to discuss the characteristics of this "proper behavior": the precepts governing the operation of the reward system of science, the "institutional norms" at once functional for achievement of the goals of

39 Storer, *Social System of Science*, p. 25.

40 S. Cotgrove and S. Box, *Science, Industry, and Society* (London: Allen & Unwin, 1970), chap. 13.

41 *Ibid.*, p. 59.

science and for maintenance of the social system of science. Robert K. Merton was the first to try to formulate the norms of scientific activity,[42] and his discussion was later elaborated upon by Bernard Barber.[43] Merton's norms were these. *Universalism* implies both that scientific contributions shall be judged solely in terms of preestablished criteria and that the scientific career shall be open to all, subject only to considerations of ability. Ascriptive factors such as color, sex, religion, and political affiliation must not bar any individual from a career in science. Similarly, when as a scientist he makes some written contribution to the discipline, these ascriptive factors should be set aside, not entering into evaluation of his work. Equally, this implies that scientists should not be rewarded because of any economic, political, or even humanitarian advantages following from their work or characterizing their motivations. *Organized scepticism* requires that contributions be judged equally and in purely intellectual terms. Neither theories nor findings must be accepted uncritically because of the eminence of their source. This norm is the abnegation of unqualified faith: it subjects all claims to truth to the criticism of the jury-room. *Communality* is the vesting of all ownership rights to scientific contributions in the scientific community as a whole. A scientist should not restrict the availability of his discovery, and though he may claim priority in discovery he may claim no more. As Merton pointed out, eponymy in science is a mnemonic and a commemorative device: it confers no special rights of use. By this criterion it is improper to seek to patent a scientific discovery, whether for financial gain or other reasons, or to keep it secret. *Disinterestedness,* finally, implies the pursuit of science not in search of prestige, power, or financial gain, but only in the cause of science itself. Both Parsons and Merton emphasized that this is not a kind of altruism: the lack of *fraud* in science is not the consequence of some higher degree of moral integrity among scientists than among ordinary mortals. It is a norm of institutional action, rooted in expediency, rather than the consequence of some psychological trait common to scientists for which there is no evidence.

Merton is at pains to point out that for the scientific enterprise to flourish, these norms must be reflected in and supported by norms and values of the environing community. Barber adds four other kinds of social values which he feels stand in the same relationship to science. These imply a norm of *rationality,* "the critical approach to *all* the phenomena of human existence in the attempt to reduce them to ever more consistent orderly and generalized forms of understanding." He opposes this commitment of mod-

---

42   Robert K. Merton, "Science and Democratic Social Structure," chap. 16 of his *Social Theory and Social Structure.*

43   Bernard Barber, *Science and the Social Order* (New York: Free Press, 1952), chap. 3.

ern Western society (which "underlies much more than science in our society, although it is most strikingly manifested there") to the traditionalism of earlier societies, committed to the acceptance of "the rule of customs." *Utilitarianism* is Barber's second norm. This implies that the prime target of rational inquiry shall be not mystical experience, or any purely mental construct, but "the empirical phenomena of everyday life." His third norm is *individualism,* "the moral preference for the dictates of individual conscience rather than for those of organized authority." The assessment of scientific truth is open to each scientist to make as he will: "Science rejects the imposition of any truth by organized and especially by non-scientific authority." A final cultural value is termed by Barber *meliorative progress:* support offered to the progressive and potentially disruptive science by a widespread belief in the need to constantly improve the condition of man's existence.

There is a final contribution to the literature on the norms of science. Hagstrom, in his discussion of the reward system of science, likens it to a system of gift exchange. Contributions are tendered in a positive desire to please the recipient, and not as a consequence of any contractual relationship. Therefore, there can be no prior necessity or obligation specifying that the exchange shall take place, or the nature of the "gift." This being the case, scientists must feel free in their choice of contributions: that is, in their choice of work. For Hagstrom, therefore, important norms of a technical kind derive from this need for independence: freedom in selection of research problems and strategies; freedom to evaluate the significance of one's results. Hagstrom suggests that scientists' conscious interpretation of "freedom in research" refers specifically to freedom in selection of problems: this is of very special importance.[44]

These properties, then, are held to characterize both the scientific enterprise itself and that ideal society best able to support and nourish it. The Mertonians believe that there is a symbiotic, mutually supportive relationship between pure science and that ideal "liberal democratic" society dedicated to precisely similar values and functioning according to similar rules of action. The Mertonian tradition is based upon the assumption that the relationship between science and sociopolitical values derives from the definition of that society best able to support the pursuit of pure science which, irrespective of social implications, most completely reflects the norms of science. The remarkable progress of American science leads Barber to identify the United States as the society most nearly approximating his ideal.[45]

It may be that there are other, preferable ways of examining the relationship between science and social and political values: I shall have some-

---

44 Hagstrom, *Scientific Community,* p. 104 *et seq.*

45 Barber, *Science and Social Order,* p. 110.

thing to say on this later. For the moment, let us accept that such relationships are best discussed in terms of the norms of science. Bearing in mind
what was said about the nature of science as a cognitive, intellectual process
at the beginning of the chapter, can we subscribe to the specifically Mertonian formulation of scientific norms? If not, then, presumably, our conception of the society ideal for science must undergo some change.

On purely philosophical grounds, I do not think we will have difficulty
retaining the normative conceptions of universalism, communality, or disinterestedness. Barber's norm of utilitarianism becomes highly problematical. Most philosophers of science would probably agree that the material
upon which scientists operate is a stuff of mental construction, certainly not
the material of day-to-day existence. Barber describes what more nearly
approximates applied science. The three norms of organized scepticism,
rationality, and individualism are also troublesome. All three seem to deny
any kind of "methodological toleration" (in Lakatos' terms), represented
for Kuhn by a rather stronger notion of commitment to the ruling paradigm.
We must recognize that science as a progressive and social activity cannot
be founded upon either total scepticism or pure individualism. And, if we
follow Kuhn's view, we may feel open to and even accept the Popperians'
charge that "our" scientists are behaving irrationally when switching from
one paradigm to another. Even under "normal" conditions, our Kuhnian
perspective indicates that the puzzles to which most scientists uniquely
address themselves are largely determined by the dictates of a voluntaristic
system of professional authority. I am not suggesting that rationality and
scepticism have no place in science, simply that Mertonians may over-
estimate their importance.

Further evidence, deriving from psychological considerations, is appropriately brought to bear upon formulation of scientific norms. I have
referred to Storer's suggestion that psychological evidence is necessarily
relevant to the problem, regarded by him as crucial, "Why should anyone
choose to undergo the prolonged education and socialization process by
which one becomes a scientist?" Equally, we may say that the institutional
norms of science must have a specific psychological relevance for the trained
scientist. In shining a psychological torch upon the scientists, we may not
unreasonably hope to observe personality traits in a sense congruent with
these norms. The findings of psychological studies on scientists' personality
characteristics indicate the following traits: high degree of autonomy and
self-direction; preference for things rather than people; preference for intellectually rather than socially challenging situations; high emotional stability;
liking for precision and exactness; dislike of personalized controversy; high
degree of control over impulses, combined with slight gregariousness;
affinity for abstract thought, combined with tolerance for cognitive ambiguity; independence of thought and rejection of group pressures toward

conformity in thinking.[46] These characteristics seem to support Mertonian norms such as individualism, rationality, and organized scepticism. Yet, as Kuhn pointed out in commenting upon these results, most psychologists have been "in search of the inventive personality, a sort of person who does emphasize divergent thinking," overemphasizing "flexibility and openmindedness . . . as the characteristics required for basic research." [47] Kuhn identifies the mode of thought appropriate to, and involved in, normal science ("even the best of it") as "highly convergent . . . based firmly upon a settled consensus acquired from scientific education and reinforced by subsequent life in the profession." (And it is notable that much of Roe's work on scientists' personalities has been based upon the study of an eminent group, nearly all members of the National Academy of Sciences.) [48] We may say, then, that psychological evidence seems to support the reasonableness of Mertonian norms within an essentially "critical-creative" view of science. If one rejects that view of the nature of science for a Kuhnian one, science's ideal society (defined by scientific norms) is less well defined than the Mertonians would have us believe: as I suggested, many social values prescribed by that tradition become problematical.

We may question the validity, or completeness, of Merton and Barber's formulation of scientific norms on entirely different grounds from those referred to above. Consider the norm of universalism. Comparison of the scientific achievements of totalitarian societies such as Nazi Germany or Stalinist Russia with those of liberal-democratic America have suggested the maleficent influence of invocation of racialist or political criteria in judging scientists' work. Such comparisons played an important part in leading Merton and Barber to formulation of the universalist norm, as well as in suggesting scientists' need for freedom from such complete authority systems. In Nazi Germany and in Stalin's Russia, to be sure, scientists had to render primary loyalty to the state. Today, however, it is questionable whether scientists regard scientific contributions, or proposed lines of advance, as evaluatable solely in scientific terms, *and* whether "liberal-democratic" society condones such "internalist" evaluations. The answer to the second question is that it does not. Powerful cultural taboos operate against, for example, performance of experiments of many kinds upon live human beings. Proper liberal sentiments may militate against execution of whole classes of scientific experiments, with the facilities representing an

---

46 Calvin W. Taylor and Frank Barron (eds.), *Scientific Creativity: Its Recognition and Development* (New York: Wiley, 1963), pp. 385–386.

47 T. S. Kuhn, "The Essential Tension: Tradition and Innovation in Scientific Research," in Taylor and Barron (eds.), *Scientific Creativity*, pp. 341–354.

48 A. Roe, "The Management of Scientists," in K. Hill (ed.), *The Management of Scientists* (Boston: Beacon Press, 1964).

inordinate drain upon public funds which might have been deployed for the greater public good. Research which seems to involve improper interference with Nature or the natural order of things is distasteful: much genetic research would be so classed. Indeed, it could be argued that the more democratic the society, the more substantial the feeling that criteria of public good should on occasion (at least) inform evaluation of scientific results by the scientific community. But is this not the case: does universalism occupy quite the place that Merton claims for it? To put the issue somewhat differently, we may question whether Merton is right in failing to include essentially "professional" norms in his list. Have no such norms emerged? Are not scientists, in some areas at least, influenced by considerations deriving from implications of their work: that is, by an awareness of their responsibility to society? Though the situation may be evolving only now, I believe that essentially professional norms (based upon appreciation of such obligations) may in some areas be attaining the status accorded by Merton to the norm of universalism. The congruence between science and democracy may be real, but different from what it has appeared to be.

A number of writers outside the Mertonian tradition have attempted to draw social and political consequences from the values and norms of science in a very different way, and with very different results. For these men, science is *least* relevantly conceptualized as extension of a body of basic, certified knowledge.

THE REJECTION OF SCIENCE

It is scarcely to be expected that rejection of science, the demonstration of its implicit authoritarianism, should have commended itself to scientists as an intellectual position. And it is not among the scientific community that the ideas of Marcuse, Habermas, Ellul, Roszak, and others with related views have found their most receptive audience. Nevertheless, such ideas are exceedingly relevant to the political sociology of science. In the first place, they have exercised an influence upon those involved with radical reform of the universities wherein the scientist carries out his work. Indeed, dean Don K. Price has suggested that the "romantic rebellion" is a more threatening challenge to science than "political reaction" (economic retrenchment), and more in need of a proper political response.[49] These two factors provide a climate of opinion within which science must continue to function. In the second place, much of the debate in the works of these men

49   Don K. Price, "Purists and Politicians," *Science,* 163 (1969), 25–31.

is concerned with the attempt to derive ethical and political implications from the norms and values of science, just as Merton and Barber have done, and, therefore, with an equal degree of relevance.

In contrast to other "prophetic" thinkers, such as Marshall McLuhan, philosophers of the new radicalism are united in rejection of something which may be called "technology" or "technics." In their diagnoses of the problem, in their views on other related concepts, and in their prescriptions for society, they have little else in common. Rejection of technology may or may not be accompanied by rejection of the whole rational mode of thought (characteristic of science); of typical democratic ideals; of political action. Critiques are formulated on the basis of perceived relationships among four factors, representable as follows:

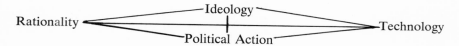

Thus their concern with the relationships between science and technology, between science and ideology, they share with many apologists for science; their concern with relationships between ideology and politics, ideology and technology (means of production) they share with many founding fathers of sociology. A second dimension useful in comprehending this multifarious philosophy, as it affects science, is the fundamental prescription offered for future salvation. Whereas, for Habermas and Marcuse, the hope lies in political action initiated by students, for Roszak it lies in replacement of the rational mode of thought, and for Norman O. Brown it lies in poetry and love. The two dimensions of these philosophies may be considered diagrammatically.

What Is Prescribed?	What Is Rejected?			
	Practice		*Concepts*	
	Technology	*Politics*	*Scientific Rationality*	*Democratic Ideals*
Political Remedy	――――Habermas―――――			
	―――――Marcuse―――――			
Intellectual Remedy	―――――Roszak―――――			
Mystical Remedy	――――――Norman O. Brown ――――――			

Roszak's book *The Making of a Counter-Culture* is interesting for analysis of the writings of not only Marcuse and Brown, but of other authors regarded as relevant to the youthful "counterculture": Paul Goodman, Allen Ginsberg, Alan Watts, and Timothy Leary. But I am concerned with Roszak's own analysis, contained in Chapter 7 of the book. The chapter title is "The Myth of Objective Consciousness." Science, the source of "objective consciousness," provides legitimation for the expertise upon which the technocracy depends. However, it is not with this function of the objective consciousness that Roszak is concerned but, rather, with science as science: objective consciousness as a way of knowing. His critique is a romantic one, not a political one (as becomes evident in Chapter 8 of his book). Scientific knowledge is objective: it [50]

> is not just feeling or speculation or subjective ruminating. It is a verifiable description of reality that exists independent of any purely personal considerations. It is true . . . real . . . dependable. It works. . . .
> What flows from this state of objective consciousness qualifies as knowledge, and nothing else does.

How absurd that this should be accepted as the only means of understanding life: [51]

> Consider the strange compulsion our biologists have to synthesize life in a test-tube—and the seriousness with which this project is taken. Every dumb beast of the earth knows without thinking once about it how to create life: it does so by seeking delight where it shines most brightly. But, the biologist argues, once we have done it in a laboratory, then we shall really know what it is all about.

Moreover, the price paid for such knowledge is complete alienation from the natural world: from the natural environment and, indeed, from that part of the self engaging in human emotion. It is an inner "shriveled-up" identity which engages in the gathering of scientific knowledge.

Roszak presents his analysis, providing a rationale for much modern eco-philosophy, as a critique of science and of scientists.[52] But what he is actually criticizing is not science (as *a* means of studying nature), but the claim that it represents the *only* means of studying nature (gathering knowledge). Scientists make this claim less frequently than do agents of the

---

50 T. Roszak, *Making of a Counter-Culture* (London: Faber paperback edition, 1971), p. 208.

51 *Ibid.*, p. 229.

52 *Ibid.*, Appendix.

technocracy on their behalf. In attempting to preserve its legitimating objective expertise, the technocrats, not the scientists, have especially derided social scientists, who can provide a "value-based expertise." What Roszak is unwittingly criticizing is *science-as-ideology:* the hegemony of the scientific way of knowing. It is to this more difficult problem that the greatly more complex thinkers Herbert Marcuse and Jürgen Habermas have addressed themselves.

In the first part of *One Dimensional Man,* Marcuse describes two ways in which technology exerts its malevolent control over postindustrial (modern Western) society. Technological imperatives determine the means of production, stipulating also that individuals' needs are directed into channels compatible with outputs of the industrial system, which caters to material, sexual, and emotional needs. Artistic and intellectual protest, as well as political protest, are neutralized by absorption: their very language is drained of meaning (the comfortable, well-equipped, super-styled *fallout shelter?*). Art, nature, eroticism, are reduced in scale so that they may be accommodated: eroticism is reduced to *Playboy* sexuality.[53] In Chapter 6, Marcuse discusses scientific rationality and its relationship to this technological totalitarianism. Is it possible to conceive of a morally neutral science, independent of such applications? No, says Marcuse:

> The science of nature develops under the *technological a priori* which projects nature as potential instrumentality, stuff of control and organization.[54]

Science as a means of apprehending nature is "hypothetical instrumentality." We are far indeed from the structural functionalist's "science as the extension of certified knowledge." The whole structure of scientific thought is now seen as logically connected with—and, indeed, based upon—the idea of domination. The principles and fundamental concepts of modern science were designed from their very beginning to serve the needs of "practical operationalism" [55] and, according to Professor Marcuse's analysis, subsequently this has come to determine the social organization of society. Scientific rationality, then, describes things in terms of their quantifiable properties: their meanings are of no interest beyond the needs of scientific hypothesis. Advanced understanding involves either replication (e.g., the creation of life) or modeling (best of all, a working model). Electronic devices can serve as models of rational behavior [56] and at the same time as instruments of control.

---

53 H. Marcuse, *One Dimensional Man* (London: Sphere, 1968), chap. 3.
54 *Ibid.,* p. 126.
55 *Ibid.,* p. 130.
56 W. Gray Walter, *The Living Brain* (London: Penguin Books, 1961), chap. 5.

Scientific rationality makes for a specific social organization precisely because it projects mere form . . . which can be bent to practically all ends.
Formalization and functionalization are *prior* to all application, the "pure
form" of a concrete societal practice. While science freed nature from
inherent ends and stripped matter of all but quantifiable qualities, society
freed men from the "natural" hierarchy of personal dependence and related them to each other in accordance with quantifiable qualities—namely
as units of abstract labour power, calculable in units of time.[57]

Though the practical totalitarianism of scientific rationality is analyzed
in detail, its *ideological* nature (at the theoretical level) is only hinted at.
It is for Jürgen Habermas to explore the social standing of scientific rationality in modern society. This he does in one of his few works available in
English. In his essay "Science and Technology as Ideology," he addresses
himself to this problem.[58]

The difficulty, which Marcuse has only obscured with the notion of the
political content of technical reason, is to determine in a categorically precise manner the meaning of the expansion of the rational form of science
and technology . . . to the proportions of a life form. . . . This is the
same process that Weber meant to designate and explain as the rationalization of society.[59]

The fundamental problem in sociological anlysis, for Habermas, is the
process by which science acquired this ideological standing. He attempts this
analysis by making the distinction between "interaction," on the one hand,
and "work" or "purposive rational action," on the other. "This institutional
framework of society consists of norms that guide (symbolic) interaction"
(p. 93). Purposive rational action characterizes subsystems "embedded"
in this institutional framework (e.g., the economic and political systems),
in which behavior is governed not by internalized social norms but by
learned technical rules. The individual's acquisition of social norms is the
development of a "personality structure"; of the technical operating rules
of a subsystem, the development of a "skill." Social systems, then, may be
distinguished by the extent to which the behavior of actors within it is governed, on the one hand, by the institutional framework, on the other, by the
subsystems: or, by the relative importance of social norms and technical
rules. Modern capitalism "can be comprehended as a mechanism which
guarantees the *permanent* expansion of subsystems of purposive rational

57 Marcuse, *One Dimensional Man*, p. 129.

58 J. Habermas, "Science and Technology as Ideology," in *Toward a Rational
Society* (London: Heinemann, 1971).

59 *Ibid.*, p. 90.

action." (p. 96). Habermas agrees with Marcuse that the ideological significance of science (its importance as a means of legitimating political power) derives from specifically modern forms of capitalism. It is perhaps not necessary to follow the complex argument in detail. A new means of legitimating political power became necessary when the Marxian analysis (applicable only "as long as politics depends on the economic base. It becomes inapplicable when the 'base' has to be comprehended as in itself a function of governmental activity" [p. 101]) lost its validity as a result of large-scale government intervention in the economy. Subsequently, politicians have chosen to focus political dialogue upon the technical problems necessary to maintain "stabilizing conditions for an economy that guards against risks to growth and guarantees social security and the chance for individual upward mobility" (p. 102), at the expense of fundamental political (moral) questions. This depoliticization of society is made acceptable by promoting science as the supreme arbiter, the source of all solutions. For Habermas, the worst is yet to come. He can envisage a transformation of the organization of society so profound that

> the institutional framework of society—which previously was rooted in a different type of action—would now, in a fundamental reversal, be absorbed by the sub-systems of purposive-rational action, which were embedded in it.[60]

But for both Habermas and Marcuse there is hope, and this hope lies in a political solution. The development of rationalization of the means of production can be a liberating influence if it serves to release individuals to take part in undominated political debate. about the future of society. Or, for Marcuse,

> ideas of liberation may become the proper end of science. But this development confronts science with the unpleasant task of becoming *political*—of recognizing scientific consciousness as political consciousness, and the scientific enterprise as political enterprise. For the transformation of values into needs, of final causes into technical possibilities . . . is an act of *liberation.*[61]

That both Marcuse and Habermas see generation of the essential conflict— by which "pacification by gratification," "depoliticization," are to be removed—as the students' task is well known. And their ideas have found a ready audience across the globe. But other radical thinkers cannot see hope

---

60 *Ibid.,* p. 106.

61 Marcuse, *One Dimensional Man,* pp. 183–184.

in politics. One such is Norman O. Brown, for Roszak the ultimate revolutionary, who pushes the psychoanalytic theory of history and of politics to its ultimate extreme. In an important exchange in *Commentary,* Marcuse rebukes Brown for his refusal to accept that symbolic interpretation can be a means to an end: that the world has to be seen as both symbol and reality: "(Brown) is stuck with the time-honored quandary of psychoanalysis: the airplane is a penis symbol, but it also gets you in a couple of hours from Berlin to Vienna." [62] Whereas for Marcuse the interpretation of events is a prelude to political action—and thence to liberation—for Brown there can be no fundamental change through politics: "The real fight is not the political fight, but to put an end to politics . . . from politics to poetry. . . . From politics to life. And therefore revolution as creation, resurrection, renaissance instead of progress. To perceive in all human culture the hidden reality of the human body." [63] Brown rejects the overthrow of technological totalitarianism not only by politics (as in the schemes of Marcuse and Habermas), but also by the transformations in perceptual schemes favored by Roszak: "One of the eternal verities is the human body as the measure of all things, including technology. The businessman does not have the last word; the real meaning of technology is its hidden relation to the human body; a symbolic or mystical relation." [64]

These radical critics of science hold, in common with one another—as with the Mertonian apologists—a belief that political implications follow from projection of the goals and values of science onto the plane of political thought. Both groups make similar kinds of intellectual projections: both, doubtless, have been influenced by empirical evidence of the unequaled fecundity of science in postwar America. Thus their differences in interpretation are attributable to their very different perspectives upon American society, a society which has proved uniquely hospitable to science. For one, the most striking factor has been the open democratic nature of American society; for the other, inequality has been more characteristic. At the same time, the two groups employ very different conceptions of what science *really is,* and of the moral imperatives governing the attainment of its institutional goals. Merton and Barber regard science as essentially creative: so (even more so) do Popper and Lakatos and most psychologists interested in scientists. Popperians dismiss the activities of the less creative toilers in the vineyard as irrelevant to science:—not science at all. In Kuhn's philoso-

62 H. Marcuse, "Critique of Norman O. Brown," *Commentary* (February 1967), pp. 71–75. (Reprinted from *Commentary* by permission; Copyright 1967 by the American Jewish Committee.)

63 Norman O. Brown, "Reply to Herbert Marcuse," *Commentary* (March 1967), pp. 83–84. (Reprinted from *Commentary,* by permission. Copyright 1967 by the American Jewish Committee.)

64 *Ibid.*

phy we find that science is regarded largely as application of preestablished procedures to consensually chosen puzzles of a much more mundane kind. Science is then vastly more circumscribed by technical rules: Kuhn takes us one step toward Marcuse. The latter, like Marx, appears to regard science as most importantly the stuff of technological innovation, an input to production, a view which has produced little impact upon Western philosophies of science. This idea is intimately bound up with that, to which many radical scientists of the 1930s subscribed, seeing progress in the field as substantially determined by technological and industrial imperatives.

Today many economists see science in this way. Jacob Schmookler, for example, collected statistical data relating inventive activity (measured by patents) and economic performance in a wide range of economic areas, and showed rather convincingly that fluctuations in the former were a consequence of economic change.[65] "One may suggest further," Schmookler argues, "that even basic inventions which establish new industries are often, and perhaps usually, induced by economic forces." Indeed, Kuhn has rebuked Joseph Ben-David for underestimating the extent to which technological factors may determine the *direction* of scientific progress.[66] Therefore, a considerable body of opinion, based upon substantial evidence, would have us regard science as determined largely (whether in rate, direction, or both) by technological and economic factors. At the same time such a perspective directs the attention toward not the epistemological results of scientific advance, but the technological results. Science *is,* most important, what results from it: innovation, technological change, whether real or merely latent. The political system most favorable to science, thus conceived, is one which can most effectively utilize the results of science and, consequently, afford the greatest investment in science. Technical norms might then be expected to direct the scientist to a concern with application of results of his work. Alternatively, such an involvement might result from moral norms, norms of a professional kind, requiring the scientist to take account of social outcomes of his work. To the extent that science is publicly regarded as basically an input to social change, such moral concern would seem appropriate: to the extent that this is so, society may rightly judge scientists by standards to which other *professionals* are expected to adhere. I have suggested that such a professional *self-image* may be emerging in the scientific community: the concern of a number of scientists and engineers with imposition of a Hippocratic Oath is surely an embarrassed attempt to compensate for the previous lack of internalized professional norms.

65   Jacob Schmookler, *Invention and Economic Growth* (Cambridge: Harvard University Press, 1966).

66   T. S. Kuhn, Review of Ben-David's *The Scientist's Role in Society, Minerva,* 10, No. 1 (1972), 166–78.

Whether for reasons of pure expediency or for moral reasons, some practical concern of scientists with the application of their work seems indicated. Robert Merton, from his very different perspective, seems to accept this in principle but to deny it in practice.[67] Yet it is so, both at the level of the individual enterprise and at the national level and, in the latter instance at least, involvement represents an essentially political activity. In the final part of this chapter, I will comment on the political actions of scientists, a theme recurring through the rest of the book.

THE FUNDAMENTAL POLITICO-SCIENTIFIC ROLES

The significance of these various attempts at relating science to political values for the analysis of political behavior lies not in their correctness but in their salience, in their influence upon the actors. If scientists believe that the pursuit of knowledge can flourish only in a democracy, the attempt to democratize society is then a legitimate, relevant action. I shall use the term *politico-scientific roles* to characterize the various forms of political action entailed by the essentially political nature of science. Both the attempt to democratize the society in which they work and the attempt to assert the autonomy of science would then seem to constitute legitimate forms of action for scientists. The latter may involve protesting against undue influence of the military establishment upon the direction taken by research; it may involve active resentment at the readiness of the National Academy of Sciences to undertake secret research for the Department of Defense; it may involve active opposition to restrictions placed upon the freedom of scientists to travel (whether Oppenheimer or Pauling in the United States or Medvedev in the USSR). These activities form the basis for one set of politico-scientific roles. It is less easy to delineate legitimate actions flowing from the attempt to democratize society, when such actions are determined by comparison of society's "real" structure with a subjective ideal. The needs of democracy are open to a wide variety of interpretations. For some, the important thing is involvement of the population in decision making. The scientists, then, perceive their role to be provision of necessary technical information in suitable form.

For others, the crucial thing is to oppose the accession to power of totalitarian figures: Stalins, Hitlers, or even Goldwaters. In this perspective, it is fascinating that one of the most effective political mobilizations of scientists witnessed in the United States was in opposition to the presidential

---

67 R. K. Merton, "Science and the Social Order," reprinted as chap. 15 of his *Social Theory and Social Structure.*

candidacy of Barry Goldwater, in the form of the organization Scientists and Engineers for Johnson-Humphrey. "Before the campaign was over, they had listed over 50,000 scientists and engineers on their membership roles; had raised some $500,000; had written and financed over 100 newspaper advertisements, 3,000 spot radio broadcasts, and a half-hour nationwide, and had elicited hundreds of columns of newspaper notices." [68] The scientists' and engineers' campaign maintained itself independent of other Democratic party groups and, having united in *opposition* to Goldwater's candidacy, dissolved after the election. Finally, there are scientists who feel true democracy cannot be achieved within the present political system and for whom, therefore, still more radical solutions are necessary.

This phenomenon is not limited to the West. In the Soviet Union, Andrei Sakharov has been calling for the democratization of the Soviet system for many years, actively protesting against the imprisonment of Soviet writers and setting up his Committee on Human Rights in 1970. And he has not been alone among Soviet scientists. We learn from his important work *Progress, Coexistence, and Intellectual Freedom* that the Academy of Sciences rejected by a substantial majority the candidacy of S. P. Trapeznikov, an extremely high-ranking neo-Stalinist bureaucrat.[69] Moreover, the extent of the spread of this "bourgeois ideology" among the Soviet scientific community has alarmed the leadership. There is talk of reeducating scientists in "Marxist-Leninist understanding" of current problems.[70]

There are two obvious objections to the ideas advanced in the preceding paragraphs, and I should like to deal with these briefly. In the first place, many data indicate the generally liberal political orientation of scientists, and various interpretations of this phenomenon have been offered.[71] It could be argued that both political liberalism and commitment to a scientific career are consequences of patterns of early socialization, involving perhaps early independence from parental authority [72] or social or religious background factors. Against this view, however, Kornhauser has argued

---

68 D. S. Greenberg, "Venture into Politics: Scientists and Engineers in the Election Campaign," *Science*, 146 (1964), 1440–1444, 1561–1563. (Copyright 1964 by the American Association for the Advancement of Science).

69 Andrei D. Sakharov, *Progress, Coexistence, and Intellectual Freedom* (London: André Deutsch, 1968), chap. 6.

70 Victor Zorza, "Soviet Scientists in Ferment," *Guardian* (October 28, 1970).

71 See, for example, P. F. Lazarsfeld and W. Thielens, *The Academic Mind* (Glencoe, Ill.: Free Press, 1958); S. M. Lipset and M. A. Schwartz, "The Politics of Professionals," in H .M. Vollmer and D. L. Mills (eds.), *Professionalization* (Englewood Cliffs, N. J.: Prentice-Hall, 1966); A. H. Halsey and M. Trow, *The British Academics* (London: Faber & Faber, 1971).

72 Cotgrove and Box, *Science, Industry, and Society*, pp. 56–57.

that if scientists' political orientations are compared with those of engineers, the former are found to be noticeably more radical—"these attitude differences between scientists and engineers cannot be attributed to differences in the two groups' social origins, religion, or parents' politics, since they were remarkably similar in respect of these factors." [73] Lipset has attempted to explain the radicalism of intellectuals in the United States (among whom he includes scientists) by means of a variety of status factors and feelings of underprivilege, and also in terms of divergence between values of intellectuals and those common in American society.[74] Other authors have placed even more importance upon the significance of intellectual and, indeed, disciplinary values. Spaulding and Turner [75] and Ladd and Lipset [76] have sought to explain the differences in political orientation of disciplinary groups of American university teachers. Though their group was relatively homogeneous in terms of socioeconomic and background statuses, Spaulding and Turner found, for example, that, whereas only 27 percent of engineers were affiliated with the Democratic party, 50 percent of botanists and 78 percent of sociologists were. Correcting for status factors, they conclude that the value systems of the specialties are critical. Ladd and Lipset delineate a similar ordering, ranging from the social sciences (most liberal) through physics, other natural sciences, engineering, to business and agriculture (most conservative). Views on specific issues, as well as voting patterns, bear this out. Thus, 54 percent of social science faculty, compared to 30 percent of chemistry faculty, approved the legalization of marijuana; 71 percent of sociologists—compared to 48 percent of physicists, 34 percent of chemists, and 24 percent of chemical engineers—expressed qualified approval of the rise of student activism. These authors, too, attribute disciplinary differences in substantial measure to professional socialization, "the post professional experiences, concerns and associations dictated by subject matter." There is solid evidence that political values are associated with the value systems of academic disciplines.

There is a second objection to my general thesis. It might be argued that organized political intervention by scientists, such as the organization Scientists and Engineers for Johnson-Humphrey is political opportunism. It could be objected, that is, that such manifestations can always be ex-

73   W. Kornhauser, *Scientists in Industry: Conflict and Accommodation,* (Berkeley: University of California Press, 1962).

74   S. M. Lipset, "American Intellectuals—Their Politics and Status," in *Political Man* (London: Heinemann, 1960).

75   C. B. Spaulding and H. A. Turner, "Political Orientation and Field of Specialization Among College Professors," *Sociology of Education,* 41, No. 3 (1968), 247–262.

76   Everett C. Ladd, Jr., and S. M. Lipset, "Politics of Academic Natural Scientists and Engineers," *Science,* 176 (1972), 1091–1100.

plained as the tactical deployment of the "prestige" of science, without recourse to any innate politicism in science such as I am claiming. In other words, scientists will form such an organization not because they are driven to it by moral conviction (deriving in some sense from professional values), but because they recognize that they will do the democratic cause more good than by campaigning individually. Probably, they analyzed the situation correctly. But why should this have been so? Why should a scientific group carry this extra weight? To respond in terms of the prestige of science is to say that people tend to attach special credence to views backed by the authority of science, even when science cannot demonstrate their truth or falsehood. This is to claim for science a power of wide legitimation. One might say that science is a set of beliefs to which people are happy to resort on occasions of doubt. This is precisely the meaning of the term *ideology,* defined by John Plamenatz as a persuasive set of beliefs "to which a community or social group ordinarily resort in situations of a certain kind." [77] Following Sorel and Pareto, Plamenatz points out that ideologies are frequently used in justification of actions. So, to plead that mass political action by scientists is no more than judicious use of the standing of science in society, is in no way to deny the thesis. It is to claim that science is an ideology and, as Plamenatz argues, although ideologies need not be political they do lend themselves to political exploitation.[78] Others, I think, would put a more strongly political interpretation upon the term. Thus, to press the ideological claims of science is to suggest an inquiry into the ways—for there must be many—in which science has been exploited in political behavior.

In practical terms, it may not always be possible to distinguish political behavior of this kind from other kinds of political action described above. In other words, united action in favor of an antitotalitarian candidate may be regarded as due either to perceived congruence of scientific and democratic values or to exploitation of science-as-ideology and, indeed, may encompass elements of both.

The clearest elaboration of the ideological nature of science is found in the work of Marcuse and Habermas. Emphasizing that the meaning of science inheres above all in its authoritarian/repressive (and, for Habermas, in particular) ideological nature, their work has given rise to a counterculture itself not lacking in political significance of a practical kind. Thus, science-as-ideology may be said to give rise not only to political roles depending upon its exploitation, but also to roles depending upon the wish to discredit it. Of course, these cannot be considered legitimate in the sense

77   J. Plamenatz, *Ideology* (London: Macmillan, 1970), p. 76.
78   *Ibid.,* p. 133.

that those deriving from maintenance of the social system of science, or congruent with the institutional goal of science, can be considered legitimate. Nevertheless, they are of fundamental importance to a systematic attempt to understand the relationship between the political and scientific systems.

chapter 3
Social Control
in Science

THE EXERCISE OF SOCIAL CONTROL

How does the social system of science work? How and by whom is recognition conferred in exchange for a contribution to science? In practice, what forms does such recognition take? In accordance with what criteria is it allocated, and what typical departures from the norms of science, as formulated by Merton, are to be observed? These are some questions with which this chapter will deal. I should perhaps take note in advance of a possible criticism. In the light of my earlier attempts at showing the dependence of Merton's formulation of scientific norms upon a questionable notion of the nature of science, the reader may be distressed to see it employed here. Justifiably, he may expect an alternative formulation based perhaps upon the Kuhnian view of science.

In this chapter, however, I am concerned principally with the exercise of control over the social behavior of scientists: with "authority" in science.

Merton allows no place for conscious exercise of legitimate authority, and a major consequence of his perspective is that many phenomena which might not otherwise be so regarded must be seen as "departures from normative behavior." But here I am involved with empirical evidence on the behavior of scientists, and the orientation in no way affects the possibility of distinguishing salient characteristics of this behavior. The interpretation then placed upon these characteristics is, however, a function of perspective, and I shall comment on the implications of adopting a different view of the cognitive nature of science toward the end of the chapter. A second consequence of adopting the Mertonian canon is that there is little place for possible systematic variations in scientists' theoretical perspectives (or "paradigms" or "research programs"), whether these exist in constant or occasional competition. Therefore, it allows us to ignore the exercise of control or authority over scientists' relations with the subject matter of their work (as distinct from their relations with other scientists).[1] The problem with which I have to deal is halved, and others must inquire into the cognitive issues which the reader may feel I am avoiding.

In this chapter I shall show that control over allocation of rewards, access to media of communication, and facilities necessary for performance of research is divided along two dimensions: between the community as a whole and an elite group within the community and between science as a whole, individual disciplines, and small specialisms within science. From such an analysis may follow an estimate of this elite's power in relation to the scientific community as a whole. (It is worth pointing out that, if one admits as a departure from the Mertonian perspective that specialisms may differ from one another both cognitively and normatively, such a division of control is essential for maintenance of the integrity of science.) Subsequently, I shall attempt to demonstrate the dependence of the criteria used in this allocation process upon those values commonly held in society-at-large. Finally, we turn to that segment of the scientific community employed in "organizational laboratories." To what extent does the jurisdiction of the scientific community and its appointed agents extend to these "organizational scientists"?

The allocation of rewards for a valued, legitimated contribution to the advancement of science is but one stage at which social control is exercised by the scientific community. The first stage lies in admission of new members to the community and control over their selection and socialization, a process which begins essentially with postgraduate study. Later, a newly inducted member, having passed through the doctoral and postdoctoral stages of his "education," needs the wherewithal to practice his trade: he needs a job; he needs research facilities. Few scientists can support themselves today and, although a minority may require no more facilities than pencil and paper, the majority need equipment of ever increasing sophistication. And,

---

1 An exception should be made for Merton's paper on the Matthew Effect, in which he discusses mechanisms by which a stratification system in science serves to focus attention upon research areas of particular excitement or fertility. See Robert K. Merton, "The Matthew Effect in Science," *Science,* 159 (1968), 56–63, below, pp. 76–77.

again, having carried out his experiments, the scientist will want to bring them to the attention of his peers—whether to secure a competent response to his feelings of creation, to receive recognition of his success, or to demonstrate the successful solution to yet another communal puzzle. Therefore, he must publish his work in a journal which he may be sure his colleagues will read: to do this, he must satisfy the editors of the meaningfulness of his contribution. And, finally, the scientific community also controls the allocation of rewards to the scientist, to the extent that he has contributed to their institutional goal: advancement of knowledge. At each stage control is exercised in an autonomous, though variously constrained, fashion by the scientific community. In the following section we shall look at some of these stages.

Access to Facilities for Research

It is very difficult to find anything approximating an autonomous allocation procedure by which research facilities are made available: external criteria and influences are insinuative here. The primary requirement is a job with adequate free time for the scientist to carry out his research. If he wishes to do basic research, motivated principally by the desire to understand and to add to the corpus of certified knowledge, in the United States this might imply a university post, in France a university or CNRS job, in the USSR a post in an academy institute. One may generalize to the extent of saying that in English-speaking countries the primary requirement is an academic post: this is the first essential for autonomous scientific research. In governmental and industrial laboratories, the possibilities of research on autonomously chosen research topics are usually not very great. Now, in the United States and Britain—though not in most European countries—university teachers are employed and appointed by individual universities. In filling tenured posts, the major universities attempt to find men who have demonstrated either outstanding research ability or research potential, whose teaching and research interests match their departments' needs, and whose character and nonprofessional tastes and orientations seem in accord with their own. The ideal process in American universities, rarely an accurate reflection of what really happens, has been described by Caplow and McGee: [2]

> The department seeking a replacement attempts to procure the services of an ideal academic man. Regardless of the rank at which he is to be hired, he must be young. He must have shown considerable research productivity, or the promise of being able to produce research. He must be a capable

2 T. Caplow and R. J. McGee, *The Academic Marketplace* (New York: Anchor Books, 1965), p. 93.

teacher with a pleasing personality which will offend neither students, deans, nor colleagues.

In working out this process, various members of the institution will be involved: to some extent the whole department, plus senior members of the administration and of other departments. However, the senior men's voices are loudest. The same would appear true in the United Kingdom, where Gaston found, in making appointments in high-energy physics, that 79 percent of professors (i.e., full professors), 24 percent of senior lecturers (i.e., associate professors), and only 4 percent of lecturers (i.e., tenured assistant professors) were involved.[3] Indeed, recently, a committee of the University of Chicago recommended that "departments, schools and committees in the university make arrangements whereby all faculty members irrespective of rank within the department possess a voice in the appointment of new members".[4] An important part of the appointive process must be to ensure making contact with suitable candidates, and Caplow and McGee have pointed out that in some specialties there may exist "one or two men who are nationally recognized as leaders. . . . Through their wide acquaintance they can place almost anyone in the specialty."[5] In general, power over appointments is spread among all the senior men in an individual discipline: all may exercise some influence over some appointments. In addition, some few leaders may exercise special influence as a consequence of their specialist prestige. If the University of Chicago committe's prescriptions approximate current practice in any sense, individual specialisms can wield little corporate power since, within the discipline as a whole, they are being judged, and should be judged: [6]

> Appointive committees, in seeking and considering candidates should, while regarding present or prospective distinction as indispensable, attend to the needs of the department in the various subfields within the discipline or subject and the *capacity of those subfields for further scientific or scholarly development.*

A suitable job is only the beginning: most scientific research costs money as well as time. Money can be used to buy apparatus, laboratory animals, computer time, assistance, or pay for a trip up the Orinoco. Research funds, for the university teacher in most Western countries, can come from three

3   J. Gaston, "Big Science in Britain" (Ph.D. thesis, Yale University, 1968), p. 338.

4   Committee on the Criteria of Academic Appointment, Report, *University of Chicago Record,* vol. 4 (December 1970). Reprinted *Minerva,* 9, No. 2 (1971), 372–390.

5   Caplow and McGee, *Academic Marketplace,* p. 99.

6   *University of Chicago Record,* my italics.

possible sources: from the university or institute, a "research council" (in the United States, NSF, bodies with responsibility for support of basic research), or a user of research who may wish to contract some of his requirements to the university specialist. In American universities and those in Europe, research is financed from all these sources, but over the past few years the balance has been shifting toward finance from nonuniversity sources.[7] This evolution must have led to a growing concentration of power in the hands of those awarding research grants at the expense of the individual university community. There are a number of things to note about this situation. Committees awarding research grants are generally organized on a disciplinary basis and, although there may be a growing "problem-orientation," power resides largely with representatives of the discipline. The multiplicity of such sources is important, partly because the greater the multiplicity the greater the likely involvement of members of the scientific community and the greater also the possible autonomy. Thus, the fact that many more agencies are funding basic research in the United States than elsewhere means the individual is less bounded by the wishes of those controlling the purse strings: he can take his application elsewhere. Although a pluralist setup suggests that many more scientists would be involved in the grant-giving procedures, very few data outline the extent of this involvement in any country. My study of chemistry teachers in British universities suggests that about 10 percent sit on grant-giving committees: undoubtedly, the figure for the United States is larger, but not likely to exceed 30 percent.[8] Nor do we know anything about the degree of openness of these disciplinary "elites."

Scientists exercise control over access to research facilities in other ways. Particularly relevant in disciplines known as "big science," in which there is a need to use large pieces of equipment usually provided for the joint use of a number of scientists, is control over time scheduling of the equipment. A major constraint in pursuit of research involving use of a linear accelerator, radio telescope, or research ship, or placing of instruments in satellites, is time and space: he who controls the utilization of such facilities exercises a profound control over development of the field. Equipment of this kind is often controlled by a director, advised by an expert committee. Groups working in the field suggest projects, which are later evaluated by the committee.[9] When such facilities are provided inter-

---

7 S. S. Blume, "The Support of Research in British Universities: the Shifting Balance of Multiple and Unitary Sources," *Minerva,* 7, No. 4 (1970), 649–667.

8 Hagstrom found that "more than one fourth" of a sample of scientists had served on committees. See W. O. Hagstrom, "Inputs, Outputs, and the Prestige of University Science Departments," *Sociology of Education,* 44 (Fall 1971), 375–397.

9 G. M. Swatez—in "The Social Organization of a University Laboratory," *Minerva,* 8 (1970), 36–58—describes how this works.

nationally, as increasingly they will be, the process is all the more compli-cated and rigorous. An example is the screening process through which an experimental proposal deriving from a British high-energy physicist must go before an experiment can be carried out on the European 28 GeV accel-erator at CERN.[10] Generally, the proposal will go first to one of the smaller national high-energy physics laboratories, where it will be discussed in an open seminar. The proposer will have to defend it against other possible competitors for time on the CERN apparatus. If it survives this stage it will be put to the CERN experimental committee, to be judged against proposals for scarce beam time deriving from other countries. A favorable verdict would lead to allocation of time. Sometimes the proposer might be invited by the scheduling committee to collaborate with another physicist suggest-ing a similar or complementary experiment. From conception to publication, the whole process can easily take three years. Control of the field by the elders of high-energy physics is exceptionally clear, the selection procedure extremely rigorous. To the extent that research in a given discipline is centralized through growing need for expensive facilities, the locus of social control must be changing. If mounting research costs increase the power of representative disciplinary committees over university communities, centra-lization of facilities must favor creation of controlling elites within specific subdisciplines utilizing these scarce facilities, at the expense of subdisciplines as a whole. Such elites are unlikely to be representative of disciplines: there will be many such in each discipline of any size.

Access to Media of Communication

Communication of findings is central to operation of the social system of science; moreover, this communication is carried out by a wide variety of mechanisms, both formal and informal. The work of Menzel, Garvey and Griffiths, and others has demonstrated how these mechanisms (published papers, conferences, the circulation of preprints, the "grapevine," etc.) necessarily complement one another. While in many disciplines the pub-lished paper is no longer the most basic means by which a scientist finds out what his colleagues are doing, it retains a central importance in the communication and social system of science. According to Price, its sig-nificance is more of a social than an informational kind.[11] Science is "public knowledge" and in his book of that title the physicist John Ziman suggests that only by publication can the scientist claim to have made his work truly

---

10   Described by M. Gibbons in a paper given at a conference on "Problems Facing University Science" (Manchester, December 1971).

11   D. J. de S. Price, in A. de Reuck and J. Knight (eds.), *Communication in Science* (London: J. & A. Churchill, 1967), p. 201.

public.[12] But publication is more than "making public": it is essential to establishment of priority (since journals keep detailed records on the date and time at which manuscripts are received); symbolically, it represents the existence of a discipline, bearing witness to the cumulative nature of scientific knowledge. Publication is the *legitimation* of a piece of work, a minimum and antecedent condition of recognition. This is so because publication in a reputable journal (the local newspaper will not do—nor will *The New York Times* or *Le Monde!*) is indicative of an elaborate process which has gone before. As Ziman puts it:

> An article in a reputable journal does not merely represent the opinions of its author, it bears the *imprimatur* of scientific authenticity, as given to it by the editor and the referees he may have consulted.[13]

Surely, it may be argued, when there are so many journals, anything can be published somewhere. Estimates put the number of current scientific periodicals at thirty-five thousand, and many new journals are on the lookout for material. The fact is, however, that journals differ vastly in their prestige.[14] There is evidence that, generally, scientists turn first to the most prestigious in their field, and that rejection here can acutely damage a young scientist's morale.[15] The high-prestige journals play a specially important part, and there may be little to gain from publishing in a periodical which is never cited (and presumably rarely read). It can easily be shown that a relatively small percentage of all scientific periodicals ensure access to the reward system of science and are especially relevant. Martyn and Gilchrist have shown that of 1,842 current British scientific periodicals only 590 were ever cited, and in the 1965 Science Citation Index 95.2 percent of citations were to only 165 periodicals [16] (15 percent were to *Nature* alone, 51 percent to only eleven periodicals). They conclude:

> If citation be admitted to be a measure of use, if use is positively related to value, and if the British Scientific journal literature is a representative sample of the world scientific literature, then most of the world's published science which is of value is to be found in a small number of journals.

---

12   J. M. Ziman, *Public Knowledge* (London: Cambridge University Press, 1968).

13   Ziman, *Public Knowledge,* p. 111 (© 1968 Cambridge University Press).

14   Whitley has taken the prestige of a journal to be a function of the number of submissions to it, and of the extent to which it solicited material. R. D. Whitley, "The Formal Communication System in Science," in P. Halmos (ed.), *Sociological Review Monograph,* No. 16 (London: 1970).

15   H. Zuckerman and R. K. Merton, "Patterns of Evaluation in Science: Institutionalization, Structure, and Functions of the Referee System," *Minerva,* 9, No. 1 (1971), 66–100.

16   J. Martyn and A. Gilchrist, *An Evaluation of British Scientific Journals* (London: Aslib,1968), p. 6.

Taking the presentation of papers to which 95 percent of citations refer as the cutoff point, they extrapolate to the conclusion that about 10 percent of the world list of periodicals—that is, three thousand and five hundred—constitute the relevant core.

In physics it has been shown that practitioners tend to place a relatively small group of journals at the top of their reading and ranking lists. For British high-energy physicists these are *Physical Review, Physical Review Letters, Physics Letters,* and *Nuovo Cimento.*[17] In the United States, the Coles found that 77 percent of all physicists read *Physical Review* "frequently" and 59 percent *Physical Review Letters.* No other journal was read frequently by more than 25 percent.[18] *Physical Review* is probably the major periodical of world physics, at least outside the USSR. Who decides what should go into the pages of journals such as this? Ziman, in a highly simplified account, attributes most of the power over publication to referees:

> The referee is the lynchpin about which the whole business of Science is pivoted. His job is simply to report, as an expert, on the value of a paper submitted to a journal. He must say whether the results claimed are of scientific interest, whether they are authenticated and made credible by sound experimental methods and good logic, and whether the paper is well expressed, not too cryptic nor too verbose, and with adequate references etc. He reports to the editor, and although this report may be sent to the author for action or rebuttal, he is protected by anonymity.[19]

This ideal scheme, editorial passivity and anonymous refereeing, fails to account for the enormous diversity of practice found not only between various journals, but even within the same journal. In a study of British social science journals, for example, Whitley found that frequently the editors exercised a very considerable degree of personal control over the content of their journals. "Over half of the journal editors interviewed made decisions on most of the submissions to their journal on the basis of their own judgment or in consultation with departmental, usually junior, colleagues."[20] Even when referees' opinions were solicited, they were not necessarily accepted unqualifiedly. Whitley found that the propensity to employ referees was a function both of characteristics of the editor (usually,

17   Gaston, "Big Science," p. 386.

18   S. Cole and J. R. Cole, "Visibility and the Structural Basis of Awareness of Scientific Research," *American Sociological Review,* 33 (1968), 397–413.

19   Ziman, *Public Knowledge,* pp. 111–112 (© 1968 Cambridge University Press.)

20   R. D. Whitley, "The Formal Communication System in Science," *Sociological Review Monograph,* No. 16 (1970).

"old prestigious editors" have regular panels of referees) and of the journal ("well established journals in 'pure' disciplines sent a higher proportion of submissions to referees than did new journals in interdisciplinary areas"). Diana Crane has found that distribution of such characteristics as academic affiliation of doctoral origins among contributors to (social science) journals reflects the characteristics of the journals' editors in some degree. Thus, editors from less prestigious universities are more likely to publish work by social scientists from other less prestigious universities. "Anonymous evaluation of articles does not change this relationship." [21]

Harriet Zuckerman and R. K. Merton carried out a detailed study of the editorial practice of *Physical Review* between 1948 and 1956. Their study allows some understanding of the complementary roles of editor and referee, and of the nature of the refereeing process: [22] much is relevant to discussion of the location and operation of social control in physics. Initially, both authors and referees are divided into high-, medium-, and low-prestige groups (on the basis, for example, of membership in honorific societies). It turned out that the two editors of the journal frequently made decisions without consulting referees. This was particularly the case with submissions from high-ranking physicists, 87 percent of whose papers were judged exclusively by the editors, compared with 58 percent from the lowest rank. Other submissions were sent to expert referees, a group of physicists "whether gauged by their own prestige, institutional affiliations or research accomplishments . . . (is) largely drawn from the scientific elite." [23] But how are the papers of authors of different rank refereed?

Four models are posed by Zuckerman and Merton: an "oligarchic model," a "populist model," an "egalitarian model," and a model based upon expertise. The first would restrict refereeing entirely to the established elite, the second would distribute it among the mass of authors, the third would limit it to peer groups of authors: thus, statuses of referees relative to authors would be overwhelmingly high in the first case, low in the second case, and equal in the third case. The actual data show no such consistency, and the authors suggest that this can be explained on the basis of their fourth model, in which referees are chosen, not on the basis of status considerations, but on that of "expertise and competence." Therefore, while referees of high status are consulted more than those more typical of the mass of physicists, the role of the latter is sufficient to invalidate the notion of an oligarchy. In one branch of physics, it has been demonstrated that

21   Diana Crane, "The Gatekeepers of Science: Some Factors Affecting the Selection of Articles for Scientific Journals," *American Sociologist,* 2 (1967), 195–201.

22   Zuckerman and Merton, "Patterns of Evaluation in Science."

23   *Ibid.,* p. 88.

30 percent of all participants had engaged in refereeing.[24] Clearly, this is one form of social control not entirely delegated to the elite. What of the division of responsibility along the dimension linking whole and subdisciplines? If Zuckerman and Merton's conclusion that referees are chosen on the basis of expertise in the area of a given paper is generally correct, this is indicative of delegation of power in terms of specialism. The editor retains the ultimate power and may, if he wishes, use it to exercise sanctions over adherents of a deviant specialism by denying them journal space. Under such circumstances the new specialism will form its own journal, and many are far more specialized than *Physical Review,* for example. Under such circumstances the editor will be a representative of a much narrower area than physics, and his referees of even narrower aspects of the field. But it is probably true that the more general journals—*Physical Review, Journal of the American Chemical Society,* etc.—enjoy particular prestige, —although there are exceptions. Thus, we may say that control over access to journal space is divided between the editor (in the case of a discipline-wide periodical, representative of the discipline) and the referees (representative of the individual specialisms). These specialisms, linking author and reviewer, may be taken as not dissimilar from the invisible colleges linking circles of co-respondents. Indirect evidence comes from the editor of *Physical Review,* setting forth his objections to the idea of anonymous submission of papers to referees: "A competent reviewer can tell at a glance where the work was done and by whom or under whose guidance." [25]

The Award of Recognition

The scientific community rewards its members' contributions in many different ways: the theoretical concept of "recognition" finds many tangible manifestations. Although the many forms have in common their use as symbols of achievement, they are awarded by different groups and in different ways. Moreover, they may have properties other than those they hold in common. The reality is as relevant to understanding the social nature of science as is the abstraction. Among the honors which scientists appear to consider appropriate indicators of their achievements we find, at one extreme, entirely informal rewards (requests for advice or invitations to address a seminar group); at the other extreme, highly formal rewards such as the Nobel Prize. Perhaps the most fundamental reward which scientists can offer one of their number is to base their work upon his theory or take up

24   M. A. Libbey and G. Zaltman, *The Role and Distribution of Written Informal Communications in Theoretical High Energy Physics* (New York: American Institute of Physics, 1967), cited by Zuckerman and Merton, "Patterns of Evaluation."

25   S. A. Goudsmit, *Physics Today,* 20 (January 1967), 12, quoted in Zuckerman and Merton, "Patterns of Evaluation," p. 87.

his area of interest. To have laid the guidelines for a new area of study or developed a new research paradigm is, in retrospect, the ultimate mark of greatness. These are the "revolutionary" achievements associated with the names of Copernicus, Einstein, Freud, Pasteur. There have been a number of studies on the origins of new scientific disciplines, and often the transposition of techniques or hypotheses is crucial. But, usually, the emphasis in such work has been upon problems of legitimation and institutionalization of the evolving field.[26] Thus, Joseph Ben-David suggests that "idea hybridization" (intellectual synthesis of ideas taken from a number of sources) is sociologically less relevant than what he calls "role hybridization." This involves "fitting the methods and techniques of the old role to the materials of the new one, with the deliberate purpose of creating a new role." [27] Partly, no doubt, because of the practical difficulties involved, although mainly resulting from the conception of science used, no one has attempted to employ the adoption of hypotheses or paradigms as basis for studying the reward system of science.[28] A very pale imitation, readily and quantifiably usable and demanding no understanding of the real operation of science, is the citation count. This measure, which has found considerable empirical favor, indicates the extent to which a given paper has been referred to by successors. Although, thus, it is indicative of the attention bestowed upon a given piece of work,[29] it has been preferred in empirical studies as a measure of quality of work.[30]

In addition to these widespread and informal forms of recognition of their appreciation of others' contributions, scientists have developed a multifarious, graded system of more formal rewards—frequently for overall achievement rather than for any specific piece of work. At the top of this scale stands the Nobel Prize, its prestige extending far beyond the scientific community and crossing all national boundaries. Of thirteen hundred U.S. physicists interrogated by the Coles, 99 percent regarded it as the highest of all scientific rewards.[31] Ranked second in the United States—and, no

26 T. N. Clark, "Institution of Innovations in Higher Education," *Administrative Science Quarterly,* 13 (1968), 1–25.

27 J. Ben-David and R. Collins, "Social Factors in the Origins of a New Science: The Case of Psychology," *American Sociological Review,* 31, No. 4 (1966), 451–465.

28 Not only are such concerns excluded by a Mertonian ontology, but it is unclear if exchange concepts of contribution and reward could be extended to cover this notion.

29 Suggested by S. Cole and J. R. Cole, "Scientific Output and Recognition" (1967) p. 388, fn. 33.

30 Cole and Cole, "Scientific Output and Recognition"; A. E. Bayer and J. Folger, "Some Correlates of a Citation Measure of Productivity," *Sociology of Education,* 39 (1966), 381.

31 Cole and Cole, "Scientific Output and Recognition."

doubt, similarly in other countries—is membership in the elite national scientific society (National Academy of Sciences, Royal Society). In addition, a wide range of medals, prizes, and honorary lectureship to which scientists may aspire are nevertheless in short supply. Among chemistry faculty in British universities, only about 8 percent have received any of these rewards.[32] One further form of rewarding scientific achievement is worthy of special note. To be asked to sit in judgment upon other scientists' work—to act as judge of their role performance—is accepted as a sign of scientific accomplishment. Tasks of this kind are essential to operation of the social system of science and may take the form of referee, editorship or membership on the editorial board of a journal, membership on a grant-awarding committee, and so on. To act as status judge is generally taken by sociologists as a form of recognition, although it may be that scientists' attitudes toward these often arduous honors are somewhat mixed. The fact is that service of this kind and, indeed, many other forms of scientific reward are more than simply *symbols* indicating previous achievement. All are different, and their place in the scheme of things can be understood only by paying cognizance to their various properties.

Honorific societies in many countries are closely connected with government. Although links assume different forms and different strengths (nowhere so tight as in Eastern European countries), the broad hypothesis is true of France, the United Kingdom, and the United States (where NAS has a special arm, NRC, to help with discharge of these responsibilties toward the federal government). These academies, for example, may be charged with representation of the country on the UNESCO-sponsored International Council of Scientific Unions (ICSU); they may be asked for advice; frequently, they are grant-aided by the government. In discharge of their honorific function, they are everywhere (including the USSR) autonomous, members electing new members from across the whole spectrum of science. In spite of their autonomy, and because of their close relations with central government, election to membership cannot be considered an abstract accolade. It must also be recognized as involvement with a real institution occupying a key place in organization of the national establishment. The second important point is that this almost supreme accolade (for it is this, too) is in the gift of a relatively small group of scientists—the present membership—which in intellectual affiliation may or may not be representative of all science.

Similar considerations must obtain in the case of membership on grant-awarding committees. On the whole, it is probably correct to regard such tasks as marks of intellectual distinction: most scientists seem more

---

32 Study by the author and Mrs. Ruth Sinclair.

than prepared to undertake such work.[33] Not all are prepared to give up the time required, however, feeling more anxious to devote themselves to their research, and indicating strict adherence to the norm of disinterestedness. Again, it is necessary to recognize that scientists' attitudes toward service on such bodies may depend not only upon their honorific function and time required, but also upon the relationships of such committees to central government; that is to say, their institutional reality is not to be ignored. (It would be of great significance to ascertain the extent to which scientists' interest in service on advisory committees is influenced by political considerations, as is the likelihood of their being asked to serve.) A second important aspect of these official committees is their hierarchic stratification. In almost any government organization concerned with financing science, the interest and scope of committees varies from the specifics of judging between individual projects to rendering advice on allocations to whole fields of science. To judge the performance of individuals working in one's own area of research is different from judging the support-worthiness of wide areas of science. In addition, whereas the first has an apparent legitimacy as reward for effective role performance, the latter has a rather different status. The difference, a matter of degree extending over a spectrum of committees, is exacerbated by the necessity of involving extrascientific criteria in high-level decision making. On these higher level committees, the individual is frequently present as *representative* of his discipline—a concept to which the sociology of science has failed to address itself. (Moreover, election to high-level committees is in reality more a reward for effective performance on lower level committees than directly for research performance.) The advisory hierarchy of science is related in a very complex fashion to the reward system of science and, indeed, to the performance of science. To treat service within it as an undifferentiated reward is valid only to a first approximation. I have said nothing about control over allocation of this "reward." Even more than with any other institutionalized form of recognition, the autonomy of science is highly constrained here, since the advisory network is an arm of the bureaucracy and so substantially subject to officials' control. Such authority as the scientific community may be said to wield is largely, and necessarily, vested in those few who for government purposes represent the whole of science. With the sole exception of France, in which disciplinary representatives on CNRS committees are elected by direct suffrage, little or no power accrues to the individual discipline here.

Very different are the less institutionalized forms of recognition: invi-

---

33 And, indeed, many claim that denial of opportunity to serve on grant-awarding committees is an important setback to their careers. The furore over the NIH blacklist of panel members is witness to this fact. See pp. 195–197.

tations to give lectures, or carry out research abroad, or to referee papers. As a rule, rewards of these kinds are conferred by members of the specialism in which the man works—not by his discipline as a whole and still less by science. Both power over allocation and distribution are far more widespread than with other more formal and lasting accolades.[34]

From considerations of scarcity, Gaston attempted to rank indicators of recognition such as those discussed above.[35] For British high-energy physicists he found that at the top of the list, apart from the Nobel Prize, was membership in the Royal Society, followed by other institutionalized rewards such as membership on editorial boards and funding/advisory committees. Less formal manifestations followed. If we take this order as obtaining, relatively invariant with respect both to place and field, we may say something about the exercise of social control over award of recognition. Control over the highest honors is vested in a small elite in each country (bodies like NAS and the Royal Society make nominations for the Nobel Prize), more representative of science as a whole than of any discipline or area of research. Rewards of slightly lower prestige tend to be in the gift of disciplinary elites: the councils of learned societies and the like, or the senior members of the high-status university departments. Informal rewards are still more widely distributed, and many participate in allocation. At this level the subdiscipline, the invisible college, may exercise some social control. In general, social control would appear to be distributed between elites and masses, between science and specialism.

The preceding discussion, on the one hand, points to stratification of the scientific community and, on the other, to power exercised by the elite in controlling the social processes of science. As Zuckerman points out, many argue that this is an ideal situation, since "the elite tends to comprise a gerontocracy which plays a major role in allocating . . . the facilities and rewards for doing science. This arrangement . . . is said to leave men still in their creative years free to work and to occupy those who have passed them." [36] Merton takes the argument that stratification, with its concomitant distribution of authority, is functional for science very much further than this. He suggests that from the flood of scientific publication scientists will tend to select for special attention papers by the more eminent

34 Of chemists in British universities, whereas only 8 percent had received any honorific award, and 20 percent served on grant-giving, editorial or advisory committees, 52 percent had received at least one invitation to make an expense-paid working trip to lecture or do research abroad in the past five years. (Unpublished results of a study by the author and Mrs. Ruth Sinclair.)

35 "If many scientists were involved in an activity, the net recognition to each . . . would be less." Gaston, "Big Science," p. 395.

36 Harriet Zuckerman, "Stratification in American Science," *Sociological Inquiry,* 40 (1970), 235–257.

authors: "Contributions made by scientists of considerable standing are the most likely to enter promptly and widely into the communication networks of science." [37] (A suggestion borne out by the subsequent empirical work of Cole.[38]) Since Merton holds that Nobel laureates (certainly among the elite) are likely to choose exciting, innovative, and fertile fields for study, the rapid preferential diffusion of *their* ideas prevents science from bogging down in trivia. Therefore, it is functional for science on substantive grounds.

Many, and especially younger scientists, would reject this view. That the exercise of authority in science by an elite should be functional, as Merton argues, seems to depend upon three assumptions. First, the scientific elite must be made up largely (as Merton tacitly assumes) of Nobelists and others of similar caliber, there purely on account of their ability. Second, these scientists must remain innovative, intellectually daring, and astute, capable of exercising intellectual leadership. Third, in fulfilling the explicit social duties delegated to them, in wielding their authority, they must not be overinfluenced by their own past histories, age, interests, and loyalties. It seems to me that none of these conditions are self-evidently met, and the issue must remain open. However, in the next section I hope to throw a little light upon the first of the three conditions. Is it solely on the basis of ability that scientists are promoted to membership in the elite? Are rewards in science based principally upon the scientific worth of individual contributions—or is the situation more accurately reflected by Ladd and Lipset's survey, in which "half of the respondents accepted . . . the charge that the most successful men in their fields gained their positions more as "operators" than as scientific achievers." [39] If this is so, Merton's view of the functionalism of stratification falls. My own view will be somewhat more moderate!

STRATIFICATION AND SCIENTIFIC NORMS

In analyzing the workings of the scientific community, it is necessary to distinguish between the institutional commitment of science to norms (such as disinterestedness, organized scepticism, rationality, etc.) and the behavior of individual scientists. The folktales of science are full of stories of behavior far removed from that prescribed for scientists. Individuals may be driven by the demands of their other roles (as husbands or fathers, for

---

37 Merton, "Matthew Effect in Science."

38 Stephen Cole, "Professional Standing and the Reception of Scientific Discoveries," *American Journal of Sociology,* 76 (1970), 286–306.

39 E. C. Ladd, Jr. and S. M. Lipset, "Politics of Academic Natural Scientists and Engineers," *Science,* 176 (1972), 1091–1100.

example) to set an "illicit" premium upon the economic rewards which they can extract for their work. They may seek a Nobel Prize rather than the answer to a problem worthy of the prize.[40] They may identify with their own theories to the extent that they are unwilling even to countenance their rejection. They may not always be entirely rational: indeed, by now it is widely accepted that scientific creativity involves processes of thinking in addition to the purely rational. (But this extrarational element will be concealed in publication of the research, indicating acceptance of the *institutional* commitment of science.)

My view is that the extrinsic social and cultural values of scientists, their other roles, and their loyalties other than to science *typically* intrude into the evaluation and control process in science. The working out of the social system of science displays many examples of such typical departure from norms: secrecy, selective citation of one's own work and that of friends, resistance to acceptance of new discoveries,[41] and so on. Perhaps the most famous anecdote among the folk literature dealing with such things is that relating how a very original paper by Lord Rayleigh (then president of the Royal Society) to which he had forgotten to affix his name was rejected—until the author's identity was uncovered. This phenomenon— the relevance of scientists' status to reception of their contributions—is far more general, and constitutes one example of *typical* departures from scientific norms.[42] Is Merton right in approving it? I have suggested that two important but sometimes forgotten features of the reward system of science are, first, that many "rewards" are rather more than abstract tokens of achievement, possessing an important institutional reality of their own and, second, that allocation of rewards is not an automated process, but represents the conscious deliberations of delegated status judges. There is no reason to suppose that these judges are likely to be either particularly lacking in external roles, commitments, or values, or especially able to set aside these externalities of their existence in making their judgments. Nor is there reason to suppose, in any *a priori* way, that they will represent the segment

40 For this reason, Professor Chauncey Leake has suggested abolition of the Nobel Prizes in science. See his letter in *Science,* 172 (1971), 1084.

41 See B. Barber, "Resistance by Scientists to Scientific Discovery," *Science,* 134 (1961), 596. Resistance may also be due to what Polanyi, in discussing the Velikovsky affair, called "tacit" judgment: the *feeling* that something may not be quite what it seems, or as worthy of acceptance. See his "The Growth of Science in Society," *Minerva,* 5, No. 4 (1967), 536. See also A. de Grazia (ed.), *The Velikovsky Affair* (New York: University Books, 1966). A more recent example of the same process of tacit rejection is to be found in high-energy physics, where experiments claiming to have found empirical evidence for the existence of quarks have been ignored or "greeted with some scepticism." See *Nature,* 232 (1971), 298.

42 Stephen Cole, "Professional Standing."

of the scientific community with which they are concerned, in terms of either internal or external values or loyalties.

> Now scientists are by definition completely disinterested persons, but the definition appears to be relaxed at just one point: just one trait of un-disinterestedness is permitted—loyalty to one's subject. That is to say if one is for example an abracadabrological scholar it is permissible, nay proper, to feel that abracadabralogy ought to receive greater recognition.[43]

Empirical studies indicate that many more considerations than this, which the reader may or may not feel is proper, intrude into the evaluative processes of science. Hargens and Hagstrom have taken association with a high-prestige American university as a reward for achievement, seeking to establish factors which may help a scientist to obtain it.[44] They were especially concerned with testing for the significance of two factors: the merit of a man's work (as represented by his productivity) and his background (as represented by the prestige of the graduate school in which he earned his Ph.D.). These authors conclude that prestige of the doctoral institution is of almost equal importance with merit of work in terms of recruitment to the faculties of top universities. Diana Crane reached similar conclusions.[45] Lewin and Duchan have demonstrated a bias against women in appointment to high-quality academic institutions. They conclude that "women faculty in the physical sciences may be evaluated on the basis of different criteria than are males when they compete for positions in academia. These criteria appear to be based on personal values and attitudes reflecting widely held socially accepted beliefs regarding the role of a woman."[46] Such bias, however, was rarely able to negate the advantages of better qualifications when a woman candidate was so favored. Naturally,

---

43 William Cooper, *The Struggles of Albert Woods* (London: Penguin Books, 1966), p. 25.

44 L. L. Hargens and W .O. Hagstrom, "Sponsored and Contest Mobility of American Academic Scientists," *Sociology of Education*, 40, No. 1 (1967), 24–38.

45 Diana Crane, "The Academic Marketplace Revisited," *American Journal of Sociology*, 75, No. 6 (1970), 953–963.

46 Arie Y. Lewin and Linda Duchan, "Women in Academia—A Study of the Hiring Decision in Departments of Physical Science," *Science*, 173 (1971), 892–895. Their conclusions follow from an inquiry based upon solicitation of departmental chairmen's responses to a set of hypothetical applicants' résumés. There is other evidence, cited by these authors, showing the same phenomenon. Women in science are not, on average, paid so well as men; they are less successful in solicitation of research grants from NSF; they are never awarded medals by the American Chemical Society. (See A. E. Bayer and H. S. Astin, in *Journal of Human Resources*, 3 (1968), 191.) This evidence, however, unlike their own, *could* be a reflection of differences in ability between the two groups, in principle. (Quotation Copyright 1971 by the American Association for the Advancement of Science.)

Hargens and Hagstrom try to explain their findings in rather different terms, which also represent intrusion of a social process common enough in society into science. Caplow and McGee had suggested that the good "contacts" made in top schools constitute an important factor in determining future career.[47] Similarly, Hargens and Hagstrom invoke Turner's distinction between "sponsored" and "contest" mobility in explaining the importance of the doctoral institution. *Contest mobility,* in which the prize (elite status) is "taken by the aspirants' own efforts" [48] is clearly the mode of ascent expected in science. And yet that there should be an element of sponsorship would seem a useful means of explaining Hargens and Hagstrom's and other data. Under *sponsored mobility,* according to Turner, recruits are selected by the established elite and elite status is *given* on the basis of some criterion of supposed merit: it cannot be *taken* by any amount of effort or strategy.[49] A second characteristic of sponsored mobility is early selection of future members of the elite, and the relevance of the prestige of doctoral institutions is evidence for this. Other aspects of the sponsored mobility process pointed out by Turner suggest its relevance for scientific careers. Turner suggests that

> individuals do not win or seize elite status, but mobility is rather a process of sponsored induction into the elite following selection.
>
> Pareto had this sort of mobility in mind when he suggested that a governing class might dispose of persons potentially dangerous to it by admitting them to elite membership, provided the recruits change character by adopting elite attitudes and interests.[50]

William Cooper's justly renowned novel of the scientific life, *The Struggles of Albert Woods,* shows vividly how this works in science. The hero, Albert Woods, working away in the unknown laboratory of a humble university at the beginning of his career, performs some interesting experiments in the chosen fields of an eminent Oxford chemist named Dibdin. Albert sends his paper to Dibdin, asking for advice, but little realizing the effect that it would produce on the other.

47 Caplow and McGee, *Academic Marketplace.*

48 Ralph H. Turner, "Sponsored and Contest Mobility and the School System," *American Sociological Review,* 25 (1960), 5.

49 *Ibid.* So far as recruitment to academically based professions is concerned, the "criterion of supposed merit" must be seen to include some element of attention to class considerations. Discrimination against the working class is apparent both at admission to postsecondary education and at subsequent stages of selection. See *Group Disparities in Educational Participation* (Background Study No. 4 for OECD Conference on Policies for Educational Growth.) (Paris: OECD, 1970).

50 Turner, "Sponsored and Contest Mobility," p. 5.

It takes a man of sublime confidence, when he sees a youngster solving a problem that he had been thinking was his own property, not to feel momentarily shaken—or if not momentarily shaken at least to feel he ought to find out exactly what the youngster is like and be quick about it.

Soon enough, Dibdin invites Albert to join his own department.

This is not the moment for me to go into Dibdin's motives. All I will observe is that if you feel a young man is likely to be a rival in your own line it is not a bad idea to send for him to work with you.[51]

Which is surely the point made by Turner and Pareto. In another paper, Diana Crane attempted to ascertain the relevance of universal and ascriptive criteria for such forms of recognition as presidency of national associations, membership on honorific societies, receipt of the Nobel Prize, of honorary degrees, and of prestigious fellowships, and membership on government advisory bodies, combined into compound indices of recognition.[52] Using an index of productivity unweighted for quality (by citations), she found that highly productive men at major universities were considerably more likely to have gained recognition of their work than were equally productive men at minor universities. Indeed, the latter were no more likely to have won recognition than very much less productive ones at the major school. Deviations from norms in allocation of rewards, such as those studied by Crane, may be understood in terms of the institutional nature of the rewards: the national association, the honorific society, the top graduate school, the government advisory committee, and so on. That is, as well as hypothetical indicators of recognition, these are institutions with their own values, traditions, and interests—which may be only partially congruent with the values, traditions, and interests of science. Thus, just as academic appointments have been denied well-qualified applicants on grounds of race, religion, politics, or sex (and, of course, personality), the same is probably true of many other forms of scientific reward. Thus, it appears that J. B. Conant (not only one of the most famous names in the history of American education, but at an earlier time a most eminent chemist) was rejected in his candidacy for presidency of NAS on grounds purely of personality.[53]

It is illustrative to consider what is known of the election process within NAS, even though I propose to postpone detailed discussion of the

51 W. Cooper, *Struggles of Albert Woods,* pp. 27–28.

52 Diana Crane, "Scientists at Major and Minor Universities: A Study of Productivity and Recognition," *American Sociological Review,* 30 (1965), 699–714.

53 D. S. Greenberg, "The National Academy of Sciences—Profile of an Institution (II)," *Science,* 156 (1967), 364.

institutional nature of such honorific bodies till Chapter 6. After the Nobel Prize, membership in the National Academy of Sciences is the highest honor to which an American scientist can aspire. The academy exercises complete autonomy in election of members, and membership is to be "in recognition of distinguished contributions to scientific and technological research." Greenberg has chronicled some curiosities of the election process as it operated in the mid-1960s and, although the numbers to be elected annually have recently been raised, there is no reason to suppose that the procedure has undergone change. In the first place, nominations are made and screened by a variety of internal committees and groups.[54] A list of acceptable nominees is submitted, together with accounts of their work, to the entire membership, some two-thirds of whom generally vote. The top seventy or so candidates are then juggled about to take account of disciplinary quotas: in 1967 eighteen life scientists, eighteen physical scientists, and three applied scientists were to be elected. (The quota for applied scientists has since been raised to ten.) The top forty-six names must now be arranged in that fashion: $18 + 18 + 10$.[55] The remaining four names, making up the annual admission of fifty, lie in the gift of the council of the academy. These fixed quotas, in conjunction with disciplinary loyalties, ensure that changes in the overall makeup of the academy are slight and that the representations of many disciplines are out of proportion to their representation in the scientific community-at-large. Thus, in 1966, although there were twice as many Ph.D. chemists as physicists (according to the National Register), their representations in the academy were virtually equal. It is for contemplation whether this is a consequence primarily of the unique contribution of physics to scientific understanding in the first half of the century, or of the loyalty and voting power of physicists. Institutional loyalty may play as much of a part as disciplinary loyalty, and Greenberg's subjective study bears out Crane's findings. A high-ranking staff member of NAS told him, for example: "Creative scientists from smaller places don't have much of a chance." Rivalry between Berkeley, Harvard, Caltech, Columbia, Chicago, and other members of the top league counted for a lot, it seemed, and those from outside stood little chance. Personality factors were also relevant: "He deserves to be in, but he's just too abrasive" was the comment upon one individual.

Recent events have illuminated other criteria which serve to determine whether or not an individual is elected to NAS. In the 1971 elections a Cornell ecologist, LaMont C. Cole, was apparently ranked forty-seventh in the juggled order, but was passed over by the council in favor of the forty-eighth

---

54   *Ibid.,* and (I) *156* (1967) p. 222.
55   "Academy Squirms a Little in Sudden Spotlight," *Nature,* 231 (1971), 6.

to fifty-first candidates.[56] The ensuing debate received more publicity than the academy is accustomed to in its internal proceedings, raising two interesting issues. Asked to justify Professor Cole's rejection, the secretary of the academy replied that he had been passed over because of "doubts about his scientific competence." It transpired that no member of the council could be termed an ecologist. So, at this elevated level, the judgment of an individual's scientific competence may be made not by, but in the total absence of, any representative of his discipline. Later in the debate it became clear that his scientific competence had not, in fact, been called into question. Opposition had derived from pronouncements on environmental issues which Professor Cole had published for popular, not scientific, consumption—"Alarmist and scientifically doubtful pronouncements about the relationship between the burning of fossil fuels and the depletion of the world's reserves of atmospheric oxygen," according to the academy spokesman. Explicitly, the president of the academy rejected the argument that because this was a popular article, and not part of his scientific work, it was irrelevant to the candidate's election. ("I said it behooves scientists to be even more sure of their facts when speaking to public bodies than when speaking to scientists who can challenge them.")

In other words, according to the president of the National Academy of Sciences, scientific role performance, as evaluated for distribution of rewards, is to be regarded as encompassing rather more than contributions to the advancement of knowledge. In particular, it is to be taken as including contributions to advancement of popular knowledge and understanding.[57] Another anomaly in award of recognition is the conferral of science's most illustrious prize (the Nobel) upon a nonmember of his country's prestige academy. While there are no comparative statistics on this, it is interesting to note that between 1950 and 1967 nine Nobel Prizes were awarded to U.S. scientists who were not members of NAS: of these, seven were subsequently elected to membership.[58] Factors such as those which we have introduced (institutional and disciplinary loyalties, nonresearch activities, personality factors) may once again be adduced in explanation.

It is perhaps less surprising that criteria other than accomplishments are relevant to scientists' appointment to government advisory committees. Once again, membership is something more (or less!) than a symbolic legitimation of scientific achievement. Again, I want to postpone detailed discus-

---

56 *Ibid.,* pp. 6–8.

57 That sociopolitical activities are general and legitimate aspects of the scientific role is the major theme of this book. We could have hoped for no more prestigious supporter!

58 Greenberg, "National Academy of Sciences (II)," p. 362.

sion of the nature of such bodies until Chapter 6, but one or two features of their organization are germane to the present discussion. First, as I pointed out, they are generally arranged in a hierarchic fashion within the apparatus of government, such that as one ascends the hierarchy their scope increases. So, too, does the relevance of extrascientific criteria to the decisions they are called upon to make. Appointment to membership at the higher levels is based upon an assumption that the appointee will share prevailing views on these nonscientific criteria. It may also reflect sheer party-political loyalties and jealousies, though this is perhaps none too common. We may say that scientific eminence is a necessary, but rarely sufficient, criterion for appointment. In the United States, the relevance of nonscientific considerations was given a good deal of publicity in furore over the NIH appointment system (see Chapter 6). The fact is, however, that scientific autonomy in the award of this form of recognition is severely constrained. Because of the real work which such committees must do in the service of government (involving promotion of the well-being of basic science, but more besides), the scientific community must share control here with bureaucrats and politicians. Thus, it is possible to regard such departures from universalism as invocation of political criteria (chronicled in the discussion of the NIH blacklist) as a consequence of this reduced autonomy and, hence, to "absolve" the scientific community from "blame." Only in the general rigidity of committee structures—reflecting small, often reluctant recognition of new specialisms—is the probably unhindered operation of the scientific community to be seen. Thus, the slow response of the advisory and grant-awarding committee system to differentiation of disciplines and formation of new specialisms constitutes a double form of negative sanction upon those dissenting from the usual view in their choice of research problems, goals, theories, and strategies. On the one hand, they are denied equal access to research funds (since applications not falling within the clear mandate of a committee are too difficult to handle), and hence restricted in pursuit of their chosen research objectives. On the other hand, they are denied what may be their due share of one specific form of reward.[59]

Therefore, a good deal of evidence points to the relevance of nonscientific criteria in evaluation of contributions to science by the scientific community. The empirical studies with a statistical bent carried out by Crane, by Hagstrom, and by others can be given depth by detailed analysis of procedures for awarding specific indications of recognition. I have tried to show how this might be done for two such honors: election to the National Academy of Sciences and appointment to a government advisory committee. In

---

59 On the response to structural change and the invocation of sanctions, see W. O. Hagstrom, *The Scientific Community* (New York: Basic Books, 1965), chap. 4, especially pp. 204–206.

other cases, slightly different forms of behavior will be found, departing in other ways from that prescribed. The award of honorary degrees, for example, may be a means of honoring not the recipient, but the institution conferring the degree: a means of linking, albeit fleetingly, a great name with one's own undistinguished institution. So, at a rather superior level, is the election of Nobel laureates to subsequent membership in NAS. These examples are representative of a phenomenon recognized by Robert K. Merton, called by him "the Matthew Effect" ("For unto every one that hath shall be given, and he shall have abundance: but from him that hath not shall be taken away even that which he hath." Matthew 25 xxix).[60] Harriet Zuckerman has shown how, even in allocation of informal recognition, the reward system of science is biased in favor of the already successful: in preferential recollection of *their* name on a jointly authored paper, in invitations to accept honorary degrees, serve on policy committees, and contribute papers.[61] The Matthew principle has some important implications: one is the inevitability, on statistical grounds, of creation of an elite. If an individual's potential for economic reward were totally dependent upon the amount of money he had available for investment, the gradual creation of an economic elite would be favored. Similarly in science, if rewards are allocated to individuals in proportion to their previous holdings, this favors establishment of an elite. If, for example, one looks at the distribution of certain honorific tasks (membership on grant-awarding committees, on editorial boards of journals, and on councils of learned societies) among university chemists in the United Kingdom, one finds evidence of this. About 20 percent perform one or more of these tasks: however, of the 260 tasks reported, one third devolved upon only 3 percent of respondents.[62]

To this elite, and to those near to attaining membership, are delegated most tasks related to evaluation of role performance by scientists-at-large, and recognition of their achievements. Governments turn to this elite in search of advice relating to support and utilization of science. On the one hand, its members are a section of the best scientists in the country,[63] likely to possess insights denied the average toiler in the vineyard. On the other hand, their prestige and visibility in the scientific community add legitimacy to decisions with which their names are associated. Thus, in addition to the tasks delegated by the scientific community, the elite finds itself asked to

60 Merton, "The Matthew Effect in Science."

61 Harriet Zuckerman, "Nobel Laureates in Science," *American Sociological Review*, 32 (1967), 390.

62 Unpublished results of a study by the author and Mrs. R. Sinclair.

63 Although, as Merton has pointed out, many of equal accomplishment may be excluded because, for example, honorific academies may have a fixed number of places. He called this the "phenomenon of the 41st chair." See Merton, "Matthew Effect in Science."

perform many tasks for government. Indeed, it is usual for government to delegate to it a share in power over development of science through decisions on allocation of resources and formulation of long-term strategy for scientific advance and production of scientific manpower. Therefore, a variety of political responsibilities seem to follow from membership in the elite.[64]

In this section I have described some factors which influence the reward of contributions to science and which, from a Mertonian perspective, are regarded as "departure from norms." Thus, a scientist's status may influence reception of his work, as may his external affiliations and statuses (including race and sex). Certain factors are particularly relevant to allocation of characteristic forms of scientific recognition and may follow from what I termed the "institutional reality" of these forms. Any alternative formulation of the nature and norms of science would substantially alter forms of behavior regarded as "deviant"—but not the forms of behavior. For example, a perspective, such as Lakatos', distinguishing between simultaneously competing research programs, would suggest a legitimate concept of loyalty. If evolving programs of special prestige were associated with specific academic institutions, some bias in favor of those institutions might appear from analysis of recognition awarded.[65]

A view of science distinguishing periods of normal science from periods in which scientists turned outside their fields in search of a new paradigm might pinpoint the occasional influence of external factors on scientists' social as well as intellectual behavior. If so, we might expect the total constellation of factors affecting evaluation of contributions to science to change over time. Finally, one might admit the possibility that various disciplines or specialisms of science differ either in adherence of their practitioners to the norms of science or in the goals, values, or norms characteristic of them. Thus, in Chapter 2 I suggested that some areas of science may have accepted certain norms of professionalism, prescribing appropriate relations to society. Under those circumstances, it might be logical to attend to external implications of a piece of research. Therefore, the first point to emphasize is that values characteristic of the environing society (for example, general conceptions of women's roles) influence the evaluative behavior of the scien-

---

64 In spite of the disadvantage of political unorthodoxy, some scientists achieve membership in the elite without in any way sharing political beliefs common to the scientific community and the bureaucracy. Therefore, there are those who are barred from some aspects of the elite's relationship with government. Rarely have there been cohesive forces sufficient to enable this "counterelite" to coalesce: a counterelite equal in its scientific accomplishment to the elite proper, but distinguished sharply from it in its political and social views. In a later chapter, I hope to show that this *has* happened in the United States.

65 For a discussion of the geography of a new "paradigm group," "network," and "cluster," see N. C. Mullins, "The Development of a Scientific Speciality," *Minerva*, 10, No. 1 (1972), 51–82.

tific community. While it is problematic whether this is an "intrusion" of external values or whether such a distinction is invalid, the practicalities of the situation are clear enough. The second point is inevitable emergence of an elite, and of its politicization by association with government.

LIMITATIONS UPON SOCIAL CONTROL

In the preceding pages, I have suggested that social control in science —over admission to the profession, access to research facilities and to media of communication, and allocation of recognition—may be regarded as distributed along two dimensions. I have sought, in a discursive way, to assess the elite's power. Now I want to examine another dimension of that power. To what extent does the authority of the scientific community (and hence that of the elite) extend to the substantial proportion of scientists employed in government and industrial laboratories? This authority is of a voluntaristic kind: the question then becomes, to what extent do "organizational scientists" choose to submit to it? To go back a stage, we may say that they will submit to the extent that they desire the reward (professional recognition) which the community can offer for achievement in their field. To what degree, then, is *this* the case?

The involvement of government—and, in Western countries, of private industry—in the support and performance of research holds important implications for autonomous operation of the social system of science, as well as for the work conditions of substantial numbers of scientists. Extra-scientific considerations of importance to these sponsors may restrict the extent to which scientists can behave as they might wish, without having to adopt strategies of accommodation to perhaps unanticipated demands. At the same time, employed scientists may be offered inducements to accept not professional recognition (in the gift of the community), but organizational recognition (status, money) in return for their work. By this means, and by its control of research facilities and its influence over choice of research problems, the organization seeks to "usurp" control of the process of science.

Thus, scientists in organizations may toil under a specific burden when organizational goals diverge from those of science itself, when research-performing institutes are concerned with something other than advancement of "public knowledge." Organizations aiming at creation of new products, development of new processes, improvements in road safety or public health: all make specific demands upon laboratories subordinate to them and insist on an appropriate commitment on the part of the laboratory. It is true that some government laboratories—some company laboratories, even—are permitted to devote part of their research effort to pure science; whether in attempting to maintain such a capability in case of major discoveries else-where, or in believing that only if such discoveries are made "in house" will they be profitable, or in feeling that this is necessary to attract talented scientists into the organization. But only in university laboratories is pure research the rule and not the exception (except in the USSR, in which most pure research is done in the Institutes of the Academy of Sciences). Most government and industrial research is carried out with some specific end in view, however remote, and any tendencies of the research staff to drift off in theoretically interesting directions must be countered, however tactfully. There are other ways in which many such laboratories differ from the "ideal." Organizational rules may involve restrictions on publication and even on internal communication; they may limit a scientist's ability to follow through an idea he has developed, they may determine the amount of time or money which he can devote to the problem at hand, they may even affect his choice of research strategy. Organizational values may orient the scientist's view of those doing other kinds of jobs; they may make him envious of the prerequisites and status which seem to accompany managerial-type work. Do scientists come into such conditions anticipating a career little different from that of contemporaries in universities, later finding themselves unable to live up to expectations, which prove very different? Do they adapt to the different conditions, adopting an orientation basically to the organization? Or perhaps scientists differ in their initial expectations—perhaps some want precisely what the organization can offer?

In discussion of these problems of the organizational scientists, empirical interest has focused mainly on industrial rather than governmental scientists, partly, no doubt, because the lack of congruence between scientific and business goals seemed particularly sharp.[66] I want to discuss something of the literature which has developed in an attempt to shed light upon implications of the system of extrascientific control for the individual scientist, and upon the ways with which he has coped.

Government Research

Excepting the prescriptive, management-oriented work of Pelz and of others concerned with improving laboratory management,[67] social scientists' interest in government scientists has concentrated on their reactions to conditions of extreme secrecy. The development of weapons utilizing atomic and nuclear processes has given rise to a substantial historical and journal-

66 S. Marcson, *The Scientist in American Industry* (New York: Harper & Row, 1960), Introduction.

67 D. C. Pelz and F. M. Andrews, *Scientists in Organizations* (New York: Wiley, 1966).

istic literature, much of it concerned with accommodation problems of specific scientists and reactions of the American scientific community.[68] From the present perspective, we may say that this literature tackles a rather simple problem. How did a substantial segment of the U.S. scientific community react to the situation in which it found itself as a result, first, of political and military exigencies (in the 1940s), exacerbated by the antirational antidemocratic climate of the 1950s? Stated thus, the problem is approachable within the framework of the sociology of science. A link with more political discussion is provided by a subsidiary question: one might inquire how and why these situations arose. On the latter point, Edward Shils has put forward an interesting view.[69] According to Shils, a major strand in the populist-fundamentalist element of American thought has been a belief in "salvationary secrecy." That is, there has long existed among a segment of American society a belief in one or more *secrets* vital to salvation of society, with continued secrecy also vital:

> The retention of the vital secret became the focus of the phantasies of apocalypse and destruction, of the battle of the children of light and the children of darkness . . . inherited from ancient and medieval dualistic heresies. . . .
> The atomic bomb was a bridge over which the phantasies ordinarily confined to restricted sections of the population . . . (e.g.) religious fundamentalism . . . entered the larger society.[70]

Here was a secret in the importance of which all could believe for the security of the realm. But this was a conviction (in the forties and early fifties) entailing recognition by the population that "salvation lay in the hands of university scientists, a species which had not previously ranked very high in their esteem" Shils, *Torment of Secrecy,* (pp. 183–184). But the reaction of the scientific community, both to development of the bomb and to the sociopolitical aftermath, a reaction triggering formation of the Federation of Atomic Scientists, has not been approached as a problem in the sociology of science: as a response, that is, to abuse of the results of science and to ferocious attacks upon the freedom of science. Today, once again, scientists feel besieged. On the one hand, many governments are turning away from the field, both because scientists have oversold the practical benefits to be expected and in response to public concern with the disamenity value of technological "progress." On the other hand, individuals are becoming conscious and resentful of restrictions which government

68 See, above all, Alice Kimball Smith, *A Peril and A Hope: The Scientists' Movement in America 1945–1947* (Chicago: University of Chicago Press, 1965).

69 E. Shils, *The Torment of Secrecy* (London: Heinemann, 1956).

70 *Ibid.*, p. 71.

laboratories may place on their autonomy—as scientists, professionals, and citizens. The communal literature of science is full of references indicative of this consciousness: scientists objecting to restrictions on free speech, shying at limitations on their right to inform the public of implications of their work, disapproving military sponsorship of research. An inquiry by Ralph Nader's Center for Responsive Law into the workings of the U.S. Food and Drug Administration found alarm among scientists employed in its laboratories on the grounds that the Agency [71]

> constantly interferes with medium and long range projects, making it almost impossible to complete them. At least part of the reason, they assert, is that certain results might embarrass the agency either by showing a lack of scientific support for an already established regulatory policy or, conversely, establishing scientific evidence of a hazard which there is no legal authority to bring under control.

It is impossible to characterize the objectives of government, to which laboratories must contribute, in as few words as have been used in this context to dispose of industry. For every one which can be criticized in the terms used above (whether true of FDA or not), another is working so effectively that scientists are as committed to organizational goals as to those of science. The fact is, though, that economic stringency, disillusionment, and other claims upon resources are almost everywhere reducing the commitment of governments to basic, autonomously pursued science. The possibility of studying government scientists within an intellectual framework contiguous with that used in discussions of the stresses and strains of industrial scientists would seem increased.

Industrial Research

The perspective employed in much earlier sociological work on scientists in industry was derived from the study of professions, and the individual was regarded as a professional (rather than *sui generis*). Kornhauser, Marcson, and Shepard are three authors able to claim that, on the whole, this represented the occupational self-image of the industrial scientist. In Strauss and Rainwater's study of members of the American Chemical Society, this perspective is spelled out:

> This is a study of chemistry as a modern profession, born of the drive toward industrialization, linked unmistakably to the latest phases of the technological revolution. . . . Primarily . . . the profession is one of

71   James S. Turner, *The Chemical Feast* (New York: Grossman, 1970). Cited in "FDA Called Unhealthy for Science and Scientists," *Nature,* 231 (1971), 277–279.

scientific discovery and the practical application of the results to the creation of new products.[72]

Scientific fields may be regarded as professionalized when they "leave" the university: when, that is, those trained in them take up employment in "client organizations." In chemistry this happened in the latter half of the nineteenth century when chemists developed professional identity: a desire for recognition as professional men.[73] This desire, this concern with public image, is a basic aspect of professionalism, and Strauss and Rainwater point out that for such reasons scientists are largely unwilling "to settle simply for the title of 'scientist' (or 'chemist' or 'psychologist')." [74] These authors, and Kornhauser,[75] are concerned with inquiring into the implications for the idea of a profession. Science differs from the old professions (medicine, law), both in its emphasis upon discovery as much as upon application and, since scientists are generally salaried employees, in the notion of "client" employed. In fact, most of what I shall say below, dealing largely with work problems of employed scientists, does not depend upon this specifically professional orientation, but is as relevantly regarded from a specifically scientific viewpoint. It is important to point out, however, that much more is entailed by the notion of professionalism: in particular, there are implications for relationships between scientists and society (the public-at-large). These relationships, in addition, involve expectations of each group toward the other: a specific kind of conduct from the scientists, public respect from society.[76]

In the work of Kornhauser and of Marcson, substantive discussion may be considered as relating largely to two sets of issues: the role of scientists in organizations, and the sources and management of strain. Adherence to the "scientific professional" role induces in scientists expectation of autonomy (e.g., in choice of research problems), need for recognition from the professional community, and dislike of bureaucratic procedures. The behavior following from strict adherence to the institutional norms of science

---

72   A. L. Strauss and L. Rainwater, *The Professional Scientist* (Chicago: Aldine, 1962), pp. 3–4.

73   The professionalization of chemistry is discussed in detail in chap. 4.

74   Strauss and Rainwater, *Professional Scientist*, p. 223.

75   W. Kornhauser, *Scientists in Industry: Conflicts and Accommodation* (Berkeley: University of California Press, 1962).

76   On status expectations implicit in the notion of professionalism, see K. Prandy, *Professional Employees* (London: Faber & Faber, 1965). The professional perspective as a basis for discussion in this area has been criticized by Kaplan, on the grounds that not only are scientists variously qualified, but many different occupational roles and role relationships exist. N. Kaplan, "Professional Scientists in Industry," *Social Problems* (1965), p. 13.

can rarely be accommodated to any great extent in the industrial laboratory. Though a measure of exploratory or basic research may be tolerated (or, on occasion, welcomed), the industrial laboratory must concern itself largely with making an effective contribution to objectives of the enterprise in the foreseeable future. As Shepard has pointed out,[77] in highly competitive industries the need to meet the immediate threat of competition may forbid provision of resources for long-term research of dubious outcome, and when the economic climate becomes more stringent any such provision is among the earliest of economies. The assumption made by these authors is that the industrial scientist must live torn between the conflicting demands of his profession and his employer, and that even those who choose a career in industrial research would (at least initially) rather spend their time pitting themselves against the vagaries of a theoretically defined nature. As Kornhauser puts it: [78]

> Professional science favors contributions to knowledge rather than to profits; high quality research rather than low cost research; long range programs rather than short-term results. . . . Industrial organization favors research services to operations and commercial development of research. These differences breed conflict of values and goals; they also engender conflicting responsibilities and struggles for power.

Scientists are supposed to be lured into the firm by the image presented before recruitment, emphasizing fundamental research, to which they hope to contribute. The very different reality soon becomes apparent.

More specifically, what are the strains and dissatisfactions under which scientists have been thought to labor, and what strategies are available (whether to management or to scientists) in order to minimize these strains? Four areas of strain may be distinguished: the selection of goals, the exercise of control, the provision of incentives, and the location of responsibility for utilization of the results of research.[79]

Generally, management seeks a substantial measure of influence over the topics upon which employed scientists work. Frequently, however, the best scientists are left pretty much to themselves. Compromise is often possible—for example, upon choice of research topic—on other matters as well. Stress has been laid upon the differences between professional (or colleague) authority and hierarchial authority. Under the first, the independent professional makes his own decisions, subject to the advice (and, ultimately, the sanctions) of the professional community; under the second,

---

77 H. A. Shepard, "Nine Dilemmas in Industrial Research," *Administrative Science Quarterly*, 1 (1956), 52–58.

78 Kornhauser, *Scientists in Industry*, p. 25.

79 Kornhauser, *Scientists in Industry*, chap. 1.

orders are passed downward. The bureaucratic taste for orderliness is likely to prefer rationalization in such areas as hours of work, control of expenditure, and provision of services. Similarly, such matters as recruitment, organization of workgroups, and assignment of work are potentially subject to nonscientific administrative control. Here, too, there is a measure of compromise—if only in the interests of efficiency—for such decisions are frequently the prerogative of the research manager. The research manager is almost always an ex-researcher, and the problem of reconciling divergent interests devolves upon him to a considerable extent. In fact, his job may differ less in kind than in the weight given nonscientific criteria from tasks delegated to the "elite" by scientific community and government. It is through his mediation that compromise upon such aspects of the work situation as choice of research problem may often prove possible. This is not the only potential informal reward. Free time may be allowed for a little private research. Attendance at conferences may be permitted or even encouraged, as may publication of good (but nonpatentable) research, in the interests of the corporate image. The research manager acts as a buffer against unbridled exercise of executive authority, and he may allow a semblance of colleague control.

Even among scientists concerned principally with building a company career, some may wish to remain in research. A major difficulty is the provision of incentives, of suitably visible forms of recognition, in place of those available outside. The firm is geared principally to provision of status and financial rewards (reflecting its own values) differing greatly from those desired by professionals. Career ladders which do not lead out of the laboratory have had to be designed. According to Marcson, a measure of freedom has constituted a principal form of less formal reward: "Involvement in substantive work decisions." But, inevitably, all this is subject to existence of a favorable economic climate.

Inquiring into the strategies which can be adopted by scientists to ease their situations, we must first recognize that they are rarely dealing with a totally intransigent management. Though it may have little relevance for the career of the average industrial scientist, the elite almost always contains a number of scientists who have made their names from an (enlightened) industrial base.[80] If the individual finds his own firm less than enlightened, he can move elsewhere. In a study of a laboratory "reoriented" from research to development, Brown and Shepherd found that many scientists expressed their dissatisfaction by taking this step.[81] But, by and large, there

[80] Wallace Carothers (du Pont), William Shockley (Bell Labs), Irving Langmuir (G.E.C.).

[81] Paula Brown and Clovis Shepherd, "The Impact of Altered Objectives: Factionalism and Organizational Change in a Research Laboratory," *Social Problems,* 34 (1956), 235–243.

is little net movement out of industrial research. The scientist can adapt to his environment: a process of "acculturation" may take place. Researchers interested in building a career in industrial research, or otherwise with the company, may develop a range of objectives consonant with what the firm can provide.[82] If he wants to stay in research, the scientist may accept the need to compromise on, for example, choice of research topic. The strictness of his adherence to the norms and values acquired in training is gradually reduced, but it would be absurd to say that he is necessarily the worse scientist for that. Technical norms remain as valid. A third alternative is that the scientist may attempt to prevail upon the company to expand the range of concessions it is prepared to make. This bargaining process may be conducted individually, or via a professional organization. The defense of the professional integrity of members has been a traditional function of such associations and it was in response to the growing employment of chemists, physicists, and so on in industry that professional scientific bodies were formed.[83] Because professionalism is said to entail a concern for the common good, the notion of "service," it may be surmised that professional orientation of scientists should increase concern over utilization of the results of research. The professional orientation should (perhaps does) lead the scientist to criticize his employers' use of his work. Recently, Ralph Nader stressed this, calling for professionals in organizations to recognize their primary obligation to society and help in public control of technological advance.[84]

The trade union is an alternative form of protective association, although its goals are usually rather different.[85] Its class, as distinct from status, orientation is presumably one factor relevant to its slight attraction to scientists and engineers in the United States. The unionization of professionals—involving at least the tacit acceptability of withdrawal of labor—is a reaction to extreme dissatisfaction plus acceptance of employee status. Writing ten years ago, Kornhauser pointed to the ineffectiveness of such a "collectivist" strategy:

> Professional unionism has been one response to the massive employment of engineers and scientists in industry, and to the relative ineffectiveness of the professional scientific society in adapting to this new condition. But unionism is primarily a method of countering one set of organizational controls (management) with another set of organizational controls. For this reason unions

82   M. Abrahamson, "The Integration of Industrial Scientists," *Administrative Science Quarterly,* 9 (1964), 208–218.

83   Kornhauser, *Scientists in Industry,* chap. 4.

84   See this book, chap. 4, pp. 106–117.

85   T. Caplow, *The Sociology of Work* (New York: Oxford University Press, 1954).

of engineers and scientists have failed to establish themselves as representatives of professionalism, and especially of professional autonomy.[86]

But this was never quite so true of Europe as of the United States. Moreover, the striking thing about scientists' relations with industry in the United States today is less their "massive employment" than their massive unemployment.[87] Some attempt at reassessing the relative claims of professional and trade union–type associations seems called for, if the relations between scientists and industrial society are to be understood.[88]

Kornhauser, Marcson, and other writers have largely rejected the collectivist strategy, stressing the need for acculturation as the preeminent mode of career management. However, there seems no need to regard these as mutually incompatible—as all-or-nothing alternatives. Adjustment to the situation, role learning, should be regarded as a rational response, which may be employed to varying extents. A balance between local and cosmopolitan orientations may be reflected in a measure of adjustment:

> A graduate aiming to maximize his chances of doing well may assess his need for autonomy as going beyond the norms allowed by his supervisor. A decision to conflict with his supervisor can be regarded as a trade of short-term recognition against the possibility of more long-term benefits.[89]

Once we allow that the participant may make a *rational* assessment of his situation, we must also allow that collective bargaining to extract concessions from management and individual accommodation may be adopted as complementary strategies.

Such analyses of the sources and management of strain are founded upon the assumption of initially complete socialization into a unique conception of the professional role, in which behavior and anticipations differ substantially from what most organizations expect and can offer. In this

---

86 W. Kornhauser, "Strains and Adjustments in Industrial Research Organizations in the United States," *Minerva*, 1, No. 1 (1962), 36.

87 Member of Sen. Edward Kennedy's staff, quoted in Deborah Shapley, "Route 128: Jobless in a Dilemma about Politics, Their Professions," *Science*, 172 (1971), 1117 (June 11, 1971; Copyright 1971 by the American Association for the Advancement of Science).

> Scientists and engineers have generally resisted any kind of organization. They see themselves as fee-for-service professionals, and it takes a long time to get away from that view. The question is whether the present situation will create the intellectual and spiritual catharsis necessary to accept the fact that they're in a buyer's market.

88 This has been attempted, but under the very different situation ten years ago, by Kenneth Prandy in *Professional Employees*.

89 B. S. Barnes, "Making Out in Industrial Research," *Science Studies*, 1, No. 2 (1971), 157–175.

context, scientists have been taken as essentially professionals (rather than *sui generis*), but bound by Mertonian norms rather than norms of service, and owing their loyalty principally to the scientific community. More recently a number of writers have denied the inevitability of such strain, although for varying reasons. First Abrahamson,[90] who accepts that academic training in science does represent a unique socialization process resulting in complete acceptance of Mertonian norms. Abrahamson suggests, however, that the young scientists, on entering industry, undergoes a "resocialization" process, resulting in partial acceptance of industrial values. Cotgrove,[91] rather differently, denies the uniqueness of the academic experience, and that those issuing from the university have been equally socialized into Mertonian norms. In his view, the socialization process which the would-be scientist undergoes prior to employment is a function of his career objectives: the influence of the university experience depends upon his intentions. He distinguishes three resulting types of scientific *identity:* the public scientist (stressing commitment to the norms of pure science), the private scientist (stressing application of science), and the organizational scientist (stressing needs of the employer). Similarly, he identifies three corresponding scientific roles. According to this perspective, strains result when identity and role do not match: many scientists will be perfectly suited to careers in industry. Empirically, it was found that public scientists in industry were most commonly subject to tension. The strategies of accommodation which they may adopt, according to Cotgrove's view, are similar to those distinguished by Kornhauser and by Marcson. Finally, Barnes,[92] whose work is based upon the empirical study of young scientists entering industry holding only bachelors' degrees, suggests that scientists have not absorbed an overriding set of values or conceptions of the scientific role: above all, they are concerned with "doing well" in their situation. Such a perspective suggests that loyalty is determined by expediency.

The work of Kornhauser and of Marcson is open to other possible criticisms. First, one might deny the validity of Mertonian norms and, while admitting the significance of a unique socialization process, claim that Merton's formulation is not correct. Such a view does not necessarily imply that the strain model is invalid. Second, one might claim that areas of science differ from one another—whether cognitively, normatively, or both. I have done this in suggesting that there may be differences in commitment to professional, or service, norms: it is equally possible that different kinds of scientists may differ in commitment to general "scientific

90   Abrahamson, "Integration of Industrial Scientists."

91   S. Cotgrove and S. Box, *Science, Industry, and Society* (London: Allen & Unwin, 1971), chap. 3.

92   Barnes, "Industrial Research," pp. 157–175.

norms." [93] Third, like Barnes, one might deny that scientists' behavior is principally regulated by scientific social norms: this might follow from a Kuhnian view of science. Mulkay has suggested that scientists' commitment is mainly to theories and methodological rules and that this commitment, rather than any social norms, prescribes their behavior.[94] Finally, earlier in this chapter, I suggested that the social behavior of scientists, *in practice,* is significantly influenced by class and status considerations and prejudices common in the society-at-large. To the extent that scientific behavior *is not* determined by scientific norms and, even more so, to the extent that it *is* determined by general societal "values" and prejudices, scientists will find themselves pretty much at home in an organizational environment embodying these values. On the other hand, to the extent to which organizational scientists are influenced either by considerations of social responsibility (true professionalism) or of class consciousness (regarding themselves as "employees"), they may find themselves subject to organizational pressures. We must recognize that the strain model of Kornhauser and Marcson may be in error not only by indicating sources of strain which do not exist (the Abrahamson, Cotgrove critique), but by ignoring sources of tension which are real.

In conclusion, a note on the interrelationship of government and industry is perhaps relevant. Studies of pressures to which scientists have been subjected—on the one hand, as a result of actions and posturings of governments; on the other, as a result of the exigencies of commercial competitiveness—have been conducted in isolation from one another. Such a distinction cannot be sustained other than on grounds of empirical expediency. First, we have to recognize the interpenetration of government and industry to which Galbraith and others have drawn attention.[95] The industrial enterprise depends upon government to ensure an adequate provision of qualified manpower (increasingly) to stabilize prices and wages and (frequently) to purchase much of its output. Indeed, in many high-

---

93 I am not using the term "professional" in the same way as Kornhauser does. Like most contemporary writers on scientists as professionals, Kornhauser is anxious to deny the relevance of "traditional" professional models, with their concern (in theory at least) to serve society. The scientist's loyalty is to his firm. I want to use the notion of professionalism to suggest the possibility that scientists may be committed to a traditional service role, in which responsibility to society ranks with commitment to the advancement of knowledge. Since most organizations cannot (or will not) allow their scientists' loyalties to society to take precedence over their loyalties to the organization, such a perspective indicates that this strictly "professional" norm may be an important source of strain.

94 M. Mulkay, "Some Aspects of Cultural Growth in the Natural Sciences," *Social Research,* vol. 36, No. 1 (1969).

95 H. L. Nieburg, *In the Name of Science* (Chicago: Quadrangle Books, 1966); J. K. Galbraith, *The New Industrial State* (London: Hamish Hamilton, 1967).

technology fields, the very nature of output is determined by government requirements. As Galbraith puts it: "No sharp line separates government from the private firm. . . . Each organization comes to accept the other's goals; each adapts the goals of the other to its own. Each organization, accordingly, is an extension of the other. The large aerospace contractor is related to the Air Force by ties that, however different superficially, are in their substance the same as those relating the Air Force to the United States government. Shared goals are the decisive link in each case." [96] Additional grounds for requiring the two phenomena (scientists' reactions to governmental and commercial exigencies) to be considered in a contiguous conceptual framework derive from admission of issues of social responsibility or class consciousness to the pantheon of scientific values.

In much of this book I shall focus on reactions of segments of the scientific community to prevailing relationships of science with government and science with industry. A higher level of analysis would discard, or drastically reduce, the assumed degree of independence of these two relationships underlying my discussion. I ask the reader especially to bear in mind the relevance of political considerations to external and internal relationships of the scientific community today.

96 Galbraith, *New Industrial State,* chap. 27, "The Industrial System and the State (II)."

chapter 4
Scientific Organization

In this chapter we will discuss the kinds of organizations into which scientists *qua* scientists form themselves, and with processes of organizational change. I shall show how scientific organizations, particularly associations of scientists, respond to, and can be significantly affected by, changes not only in the structure of scientific knowledge but in socioeconomic structure of the environing community.

It is possible to distinguish three types of relevant function performed by associations of scientists. Their emphasis by specific organizations may be taken to reflect the predominant orientation of members toward the scientific role. In Chapter 3 three such orientations were differentiated: "scientist" (Cotgrove's "public scientist"), "professional" (Cotgrove's "private scientist"), and "employee" (called by Cotgrove "organizational scientist"). Thus, scientists *qua* scientists (whether scientist, professional, or employee) will be interested in joining associations helping them perform that (scientific) role which they regard themselves as filling. Bodies concerned essentially with facilitation of basic science, and which would attract the scientist to membership, we call scientific or learned societies. Those specially concerned with bolstering the professional or employee variants upon the scientific role we shall term professional associations in the first instance, trade unions in the second. The distinction, however, is an analytical one: often, a specific organization interests itself in more than one of these functions. It is now generally accepted therefore, that a kind of spectrum divides professional bodies from trade unions.[1] And the American Chemical Society (ACS) is both learned society and professional association. This chapter is concerned with organizational change deriving from changes in functional emphasis, as well as with that deriving from developments in the structure of science but not involving change in function for the organization. The distinction may be more clear when couched in less abstract terms.

The term "scientific society" is generally used to refer to scientists' associations motivated primarily by the goal of furthering science (or a science). The Royal Society (RS), founded in London in 1660, set down its objectives as follows:

> The business of the Society in ordinary meetings shall be to order, take account, consider and discourse of philosophical experiments and observations; to read, here [sic], and discourse upon letters, reports, and other papers containing philosophical matters; and also to view and discourse upon rarities of nature and art; and thereupon to consider what may be deduced from them or any of them; and how far they or any of them may be improved for use or discovery.[2]

Members pursued a regular correspondence with foreign societies and with corresponding members abroad and these letters, together with the record of meetings, formed the *Philosophical Transactions* of the Royal Society (a publication with a continued existence since 1665). Thus, at once this journal was a means of publishing scientific results and conducting scientific controversies. Publication of journals, so crucial to the scientific process, is still a major function of scientific societies: in the United States some 60 percent of scientific journals are published by the learned societies, although the United Kingdom figure is only about 40 percent. Some sixty years after foundation of the Royal Society, the first comparable American society of note was founded by Benjamin Franklin in Philadelphia. In 1743 this became the American Philosophical Society.[3] There followed a proliferation of societies across an increasingly wealthy America. Like their European counterparts, these organizations—there were twenty-six by

1 For the concept of "unionateness," see R. M. Blackburn, *Union Character and Social Class* (London: Batsford, 1967), chap. 1.

2 First journal book of the Royal Society, quoted in M. Ornstein, *The Role of Scientific Societies in the Seventeenth Century* (Chicago: University of Chicago Press, 1938), p. 108.

3 Ralph S. Bates, *Scientific Societies in the United States* (New York: Columbia University Press, 1958), p. 5.

1835 [4]—were involved with promotion of science and of interest in science. Though today the majority of scientific societies limit themselves to a very restricted area (a single discipline or a subdiscipline), their function with respect to that discipline is little different from what their predecessors attempted to carry out for science as a whole. They are involved with maintenance of channels of communication and promulgation of interest.

"Professional associations" are rather different. It is not necessary to enter into the somewhat vexed questions of "what constitutes a profession" or of the significance of growth of new science-based professions for the theoretical notion of professionalism. Millerson has divided professional associations into four categories (four analytic types): prestige, study, qualifying, and occupational associations.[5] The latter two most particularly accord with the generally current notion of professional association. The qualifying association emphasizes training, being concerned with legitimation of an individual's right to practice, and thus certifying that he has been properly trained. The occupational association is involved especially with protection of members' interests (whether individually or collectively): the American and British medical associations are examples. Frequently, an individual body will both qualify and protect, and a brief definition would be something like this:

> A professional association is primarily concerned with status. There is an emphasis on qualification, bestowed by the association itself or by some educational body, which may be either an indication of competence, or, in a more developed form, a license to practice. Basically the intention is the same—to guarantee to the prospective client (using the word to include employers) that the practitioner for whose services he is paying does have at least a minimum level of competence.[6]

"Trade unions" of scientists are relatively rare in both the United States and the United Kingdom. Traditionally, unions are occupational associations of blue-collar workers (although in Great Britain the early unions of the 1850s preceding the mass unionization of unskilled workers, such as the Amalgamated Society of Engineers, were limited to skilled artisans.) [7] Professional men and modern "white-collar" workers in general have tended to avoid unionization, generally taken as an indication of

---

4   *Ibid.,* p. 51.

5   G. Millerson, *The Qualifying Associations* (London: Routledge & Kegan Paul, 1964).

6   K. Prandy, *Professional Employees* (London: Faber & Faber, 1965), p. 65.

7   H. Pelling, *A History of British Trade Unionism* (London: Penguin Books, 1963), chap. 3.

identification with the working class.[8] Trade unions are protective: a major function is collective bargaining with employers on behalf of the membership over wage rates, demarcation (who does what?), work conditions. This may or may not constitute the limit of the union's activities. Lipset has called unions which do nothing more than this "business unions." [9] Others, holding what he calls "diffuse political ideologies," are more concerned with representation of class interest in the political sphere—although this is perhaps less common in the United States and the United Kingdom than in continental Europe. Collective bargaining, the essence of union work, lays stress on group advancement. Therefore, it appeals to those for whom the economic organization of society offers little hope of individual advancement. White-collar workers and professional men in particular (including employed professionals) have tended to adopt a more individualistic line, seeking *self*-reliance. Nevertheless, white-collar unionism is now growing rather substantially in many countries, and the editor of a collection of national case studies has recently written:

> Experience thus leaves little doubt that many, if not most, white-collar groups can be effectively organized. Even the high skill professional group are no exception to the statement.[10]

Therefore, while unionism has had little appeal for scientists, the possibility of its widespread adoption would seem at least a plausible hypothesis.

These, then, are the three types of organization with which this chapter, and to some extent the following one, are concerned. But I am also interested in the process of change and change is best illustrated by a little history since, though observational inquiry may allow study of organizational change "at the margin," it is essentially a historical phenomenon. The next section discusses briefly the development of chemical organization in the United Kingdom and the United States. Of all sciences, chemistry was most rapidly "professionalized," and in this area relevant phenomena were best developed.

CHEMICAL ORGANIZATION

In simplified terms, modern chemistry dates from the end of the eighteenth century, founded upon the work of Antoine Lavoisier and John Dalton. In 1772 Lavoisier realized that when metals and some other substances underwent combustion or calcination the essential process, leading to an

8   Prandy, *Professional Employees,* chap. 2.

9   S. M. Lipset, "The Political Process in Trade Unions," *Political Man* (London: Heinemann, 1963), chap. 12.

10   A. Sturmthal (ed.), *White Collar Trade Unions* (Urbana: University of Illinois Press, 1967), p. 397.

increase in their weight, involved combination with air. When Lavoisier heard of Joseph Priestley's isolation of what he (Priestley) had called "dephlogisticated air" by heating mercuric oxide, he began to appreciate the true nature of these processes. He suggested that combustion was not loss of the mysterious entity "phlogiston," as most chemists then believed, but was a combination with "oxygen" (dephlogisticated air) from the air. Lavoisier also gave us our modern system of nomenclature, replacing with it many of the older, fanciful names ("powder of algaroth," "flowers of zinc"). He distinguished between simple and compound substances, and between the various processes observed (physical changes, changes of state, chemical changes). It has been suggested that Lavoisier, by demonstrating the applicability of the Newtonian concept of mass to chemical theory, "put physics at the heart of chemistry." [11] At the beginning of the nineteenth century, Joseph-Louis Proust showed, in the face of controversy, that "when elements unite together to form chemical compounds they do so in certain fixed proportions," thereby laying the foundations for Dalton's atomic theory. Dalton's assumptions, first published in 1807, may be summarized as follows. Every element is composed of atoms which are never subdivided during chemical reactions. The atoms of any given element are exactly alike and constant in weight, but atoms of different elements differ in weight and other properties. Chemical compounds are formed by the union of the atoms of different elements in fixed integral numerical proportions. Slowly, chemistry became systematized, and at the same time it became apparent that many of its principles applied equally to animal and vegetable (i.e., living) matter. By 1814 Berzelius was able to demonstrate that many "organic" materials, obtained from living tissue, obeyed Dalton's laws of combination. In 1828 an important breakthrough occurred when Wohler synthesized a substance named urea (known to be made by the kidney) in the laboratory. Thus was born what we now call "organic" chemistry, although the term is no longer strictly appropriate, since the compounds are in no sense living.

The rapid, impressive development of the chemical industry in the nineteenth century was to a substantial extent a consequence of harnessing chemical science to technology. This was particularly the case with organic chemistry and the organic chemical industry. The story of how chemical interest in the substances preparable from coal tar led W. H. Perkin to attempt to make quinine from the coal tar derivative naphthalene, and his subsequent and accidental discovery of the aniline dyes is well known.[12]

11 S. Toulmin and J. Goodfield, *The Architecture of Matter* (London: Hutchinson, 1962), p. 228.

12 D. W. F. Hardie and J. D. Pratt, *A History of the Modern British Chemical Industry* (Oxford: Pergamon, 1966), pp. 64–67.

Henceforth, the dyestuff industry developed hand in hand with chemical research. The pharmaceutical industry, too, was founded upon organic chemistry, although it was late in the nineteenth century before traditional recipes began to be supplanted. In 1842 the great German chemist Liebig published his *Organic Chemistry in its Application to Physiology and Pathology*. In the 1880s the synthetic drug industry developed, notably in Germany, and in 1888 phenacetin was prepared, in 1898 aspirin.[13] The same Liebig was also the founding father of agricultural chemistry and the fertilizer industry. Other branches of the chemical industry based upon utilization of other chemicals—sulphuric acid, for example—were developing at the same time.

Although chemical organization as we know it today dates from the middle of the nineteenth century, ephemeral chemical societies are known to have flourished toward the end of the eighteenth. In London a chemical society appears to have been meeting fortnightly in a coffeehouse by 1781, and one sponsored by Joseph Black met in Edinburgh around 1785. Similar bodies existed in the United States, but their transient nature is indicative of the lack of real need. After Lavoisier and Dalton, the intellectual respectability of chemistry was established and it became increasingly reasonable for men to identify themselves with the subject. The first national chemical society, which still exists today, was founded in London in 1841. It was followed by the Societé Chimique de France in 1857.

Under the presidency of Sir Joseph Banks, between 1778 and 1820, the Royal Society had been a powerfully conservative force in British science. It resisted potently, although unsuccessfully, the attempts of geologists (in 1807) and of zoologists (in 1826) to break away and form their own societies—partly in response to its own ineffectiveness. By 1840 a precedent had been established, and the founders of the Chemical Society encountered no opposition from the Royal Society. The moving spirit behind its foundation was Robert Warrington, who organized a preliminary canvass of interest and then the meeting at which it was agreed to form the society.[14] A provisional committee was formed and, when its report was presented to the first meeting of the society, seventy-seven original members had signified their support. Some of those invited refused to join, however, either because they considered chemistry an inadequate base for the formation of a society, or because, as medical men, they felt that "chemical pursuits injure medical practitioners in the eyes of the public." The objectives of the society were to publish communications on chemical discoveries and observations; to build up a library; and to secure the establishment of a

13   *Ibid.*, pp. 74–76.
14   *The Chemical Society 1841–1941* (London: Chemical Society, 1941), pp. 11–13.

chemical laboratory. In 1848 the society was incorporated by Royal Charter as a body open to all with an interest in chemistry:

> For the general advancement of chemical science, as intimately connected with the prosperity of the manufactures of the United Kingdom.[15]

In 1845 the prestige of chemistry in the United Kingdom was vastly raised by founding the Royal College of Chemistry, for the first time offering "professional chemical training." [16] To a considerable extent, founding of the college was due to the prestige of Liebig's work at the University of Giessen. Liebig had developed a science of fertilizers, demonstrating the relevance of chemistry for agricultural practice. His work had produced a substantial impression on the wealthy British land-owning class, and the college received widespread, distinguished backing. To the directorship of the college was appointed a protégé of Liebig's, A. W. von Hofmann. Although the Chemical Society had not been instrumental in founding the college, it profited from the augmented status of the discipline and in addition, the founding represented fulfillment of one of its initial objectives. It concentrated on developing its publications: the occasional *Memoirs and Proceedings* published between 1841 and 1848 was replaced by a *Journal* appearing quarterly, and from 1862 monthly.

By 1867 a discontent had arisen with the management of the society, and two distinct views had developed among members as to its role. Some held that the Chemical Society should establish itself as an association of eminent men of science, and that the fellowship should be conferred only upon those who had given evidence of training and competence in chemistry. Others held a contrary view: the charter required that the society exist for promotion of interest in chemistry, and this could best be done by attracting to membership all with an interest in the subject. At the time the dispute was inconclusive, mixed up with another about the concentration of power over the society's affairs in the hands of a small London elite.

Through the 1860s and 1870s, interest in chemistry was at once growing and changing. Changing, because the period witnessed a burgeoning employment of chemists *as* chemists, not only in a growing university system, but in a developing industry. In 1851 Owens College (Manchester University) was founded and from the outset had developed a concern for science. At the end of that decade, London University had introduced its B.Sc. degree. At the same time, the dyestuffs industry had been expanding

---

15   *Ibid.*, p. 21.

16   D. S. L. Cardwell, *The Organization of Science in England* (London: Heinemann, 1957), pp. 66–77.

since Perkin's discovery of 1856. Chemists began to gain employment in the chemical industry, in brewing, in the extractive industries (e.g., mining), and in teaching. Professional training and professional careers became available to chemists. In 1875 the "Sale of Food and Drugs" Act required appointment of public analysts to check on the purity of food and drugs offered the public and these, too, were chemists. A considerable volume of criticism of professional chemists followed, directed at the quality of those employed, both as analysts and as industrial chemists. Unlike the German chemical industry, which paid its scientists on a handsome scale, the British industry offered very poor salaries—but to very poor quality manpower, it seems. The industrialist Ivan Levinstein, educated in the famous chemical laboratories of Berlin, gave expression to such an opinion: [17]

> If we can turn out of our universities and colleges not mere testers, but chemists with a superior knowledge of the hydrocarbon derivatives and especially of the coal tar colours . . . such chemists will not only be far more useful in promoting the advancement of these arts, but they will also readily find far more remunerative employment than our present chemists.

Nevertheless, employment was on a growing scale, and this was to pose an important challenge to the Chemical Society. In 1876 an article appeared "On the Necessity for Organization among Chemists for the Purpose of Enhancing Their Professional Status" in the periodical *Chemical News* (founded in 1859 and edited by the well-known scientist William Crookes).[18] The author, Dr. C. R. Alder Wright, F.R.S., proposed formation of an association, or guild, which would license qualified chemists in the way that doctors were licensed to practice. The suggestion was warmly received by a section of the Chemical Society membership, and a "professional organization committee" was formed to hold discussions with the society's council.

> With the view of ascertaining how far that Society might be able and willing to carry out a scheme for the organization of professional chemists.

The "professional chemists" were anxious to establish both their superior status within the ranks of the Chemical Society over amateur members, and

---

17 In a lecture given in 1883. Quoted by L. F. Haber, *The Chemical Industry During the Nineteenth Century* (London: Oxford University Press, 1958), p. 190. Haber contrasts the inadequacy of British chemists with the high quality of those employed in Germany, reflected in differential rewards. Moreover, the German chemists "knew that their services were appreciated. . . . Consequently the number of industrial chemists grew and they formed a powerful professional body, dominating one of the chief German industries and influencing its policy." Haber doubted whether high-quality scientists would have enjoyed a similar career in the British industry. See "Employment in the Chemical Industry."

18 *Chemical Society 1841–1941*, p. 51.

their status within society-at-large. The latter was far from high. Could the society contain this dissident faction and fulfill the requirements of both groups of chemists? The original proposal of the "organization committee" was for an Institute of Professional Chemists within the CS.[19] Fellowship of the latter would have been a necessary prerequisite of licensing by the former. Legal advice suggested, however, that the society lacked the right, under its charter, to make a special select body among its fellows. Thus, the organization committee decided to proceed independently.

In 1875 the public analysts had formed themselves into a Society of Public Analysts, but the proposal that this form the nucleus of a new body to include all professional chemists found little favor. Thus, in 1878 the organizing committee applied to the government for permission to incorporate an Institution of Professional Chemists.[20] Because of common usage of the term "chemist" to include pharmacists or druggists, the government, somewhat nonplused, sought advice from the by-then-well-established Pharmaceutical Society on appropriateness of the title. The pharmacists objected on grounds that they too were professional chemists. A related difficulty arose from the fact that the Pharmaceutical Society had acquired the right (and, indeed, the state-imposed obligation) to license "chemists and druggists." Thus, it became difficult for the newcomers to formulate articles of association empowering them to hold qualifying examinations central to their purpose. By 1879 these difficulties had been resolved. An Institute of Chemists ("Professional" having been dropped from the name) had been registered, and granted permission to hold examinations and award certificates of competence. The Chemical Society and the Institute of Chemists coexisted in harmony: the former as a study association, or learned society, the latter as a professional association. The former was (and is) concerned with the advancement of chemical science, the latter with the status of chemistry and chemists. Although, obviously, overlap in membership has been substantial, the institute appeared less relevant to academic chemists.[21] For the large numbers of chemists employed outside the pure

---

19 Curiously, the merger taking place at the end of 1971 has produced exactly this situation—over one hundred years later!

20 R. B. Pilcher, *History of the Institute of Chemistry 1877–1914* (London: Institute of Chemistry, 1914).

21 Who were "temperamentally adverse to the activities of professional societies." A. M. Carr-Saunders and P. A. Wilson, *The Professions* (London: Oxford University Press, 1933), p. 170. This, if true, represents a split between the discovery- and application-oriented ends of the profession, which must be negative for a knowledge-based profession. Such a situation is not unknown in medicine. As Sir George Pickering, F.R.S., ex-Regius Professor of Medicine in the University of Oxford, has said:

> Intellectual nihilism is the very stuff of which scientists are made, but it is scarcely convenient for a practicing physician.

Quoted by P. Ferris, *The Doctors* (London: Penguin, 1967), p. 84.

research laboratory, however, the need to regulate relationships with clients and employers assumed considerable importance. Only by distinguishing themselves from the unqualified by the use of "letters" (e.g., Fellows would write F.I.C. after their names) was it felt that public confidence in chemistry could be maintained.

Although chemistry in America developed under very different socio-economic conditions in the nineteenth century, it generated similar needs and similar organizations to meet them. In America, too, ephemeral chemical societies had existed from the end of the eighteenth century. (These included the Chemical Society of Philadelphia, 1792–1803, with which Joseph Priestley was associated.) [22] The movement to establish a national society dates from 1874. In April of that year Dr. H. C. Bolton wrote in the periodical *American Chemist* that, since 1774 had seen the "birth of modern chemistry" (it was the year in which Priestley had discovered "dephlogisti-cated air" or oxygen), 1874 was its centenary. This "Centennial of Chemistry" was worth celebrating.[23]

> It would be an agreeable event if American chemists should meet on the first day of August 1874, at some pleasant watering place, to discuss chemical questions, especially the wonderfully rapid progress of chemical science in the past one hundred years.

Subsequent correspondents were much in favor, and one (Miss Rachel L. Bodley, Professor of Chemistry at the Women's Medical College, Pennsylvania) felt it would be appropriate if the meeting were held at Northumberland, Pennsylvania, Priestley's burial place. At the May 11 meeting of the New York Lyceum of Natural History, Dr. Bolton proposed, and it was agreed, that "a social reunion of American chemists for the mutual exchange of ideas and observations would promote good fellowship in the brotherhood of chemists." A committee of five was appointed to make necessary arrangements, and the reunion took place at the beginning of August 1874, in Northumberland, Pennsylvania. Eighty chemists attended. At the reunion Professor Persifor Frazer proposed establishment of a chemical society. The suggestion did not receive universal acclaim. It was opposed by Professor J. Lawrence Smith, on the grounds that the American Association for the Advancement of Science (AAAS) [24] and the American Academy of Arts and Sciences catered perfectly well to the needs of chemistry.

It was agreed that a committee should be established to discuss with

---

22　Bates, *Scientific Societies*, pp. 52–53.

23　Marston T. Bogert, "American Chemical Societies," *Journal of the American Chemical Society*, 30 (1908), 163–182.

24　Based to some extent upon the British Association for the Advancement of Science (itself a reaction against the decline of the Royal Society, and founded in 1831), AAAS first met in 1848. Its objectives were "by periodical and migratory meetings, to

AAAS formation of a chemical section within the latter: hitherto, chemistry had been grouped either with physics or materia medica and there had been relatively little chemical discussion at meetings. Formation later that year of a discrete subsection did not satisfy the militants. In January 1876 a meeting was held in New York City to arrange for establishment of what was to have been a local New York chemical society. A circular was sent to about one hundred chemists in the area. So great was the response, however, that it was resolved to try to form a national society. Sixty chemists outside New York wished to join. An organization meeting was held at the New York College of Pharmacy in April 1876. Dr. John W. Draper was elected president, and the first meeting was held in May. Some still considered the society unnecessary and ill advised. Nevertheless, arrangements were made with the *American Chemist* to publish the proceedings, and the *Journal* was founded in 1879. The period 1880–1888 was hard for the new organization even though (or because) chemistry received the recognition from AAAS that it had been seeking. By 1888 the society "appeared almost moribund."

A major difficulty inhered in the fact that, although the society purported to be national in scope, it had been incorporated in New York, and all meetings were held there. New York members became disenchanted with carrying the whole burden of responsibility, while those outside the city were unable to attend meetings except with a great deal of inconvenience. So vast was the dissatisfaction that, at the 1888 meeting of AAAS, it was resolved to look into formation of a truly national body. The problem was to be faced by all national associations: how to permit widespread participation over such a large area. In 1889 the American Chemical Society (as the new body had been called) came up with what was to become a popular solution. In that year the constitution was amended to permit formation of local sections and to require that semiannual meetings be migratory. The local sections would have their own officers, meetings, social events, and so on. In this way members far from New York would be able to participate actively in running the society and its events, and all would gain more from membership than receipt of a journal which could be consulted in the library anyway.[25] The first national meeting outside New York was held at

---

promote intercourse between those who are cultivating science in different parts of the United States; to give a stronger and more general impulse, and a more systematic direction to scientific research in our country; and to procure for the labours of scientific men, increased facilities and wider usefulness" (Bates, *Scientific Societies,* p. 75). Meetings had always recognized the importance of specialist interest, and were held in specialist sections.

25 The American Medical Association evolved in the opposite direction, starting as an annual meeting of delegates from state associations. There was a progressive delegation of power to the center.

Newport (R.I.) in 1890, and a Rhode Island section chartered in 1891. Subsequently, a number of preexisting local chemical societies became incorporated as sections of ACS, swelling the membership of the parent body. In a sense, devolution was symbolically completed when the New York membership was forced to register as a local section.[26]

Having successfully coped with the geographic problems, the society would have to face other difficulties. First a word about eligibility and membership, an issue which, it will be recalled, posed serious difficulties for the Chemical Society in Britain. Unlike this latter body, open to all who were interested, ACS restricted membership: "Any person who has had adequate training in chemistry may be nominated for election as a member." In 1902 this was changed, and for the above clause was substituted in the statutes, "Any person interested in the promotion of chemistry may be nominated for election as a member," making it a purely learned society (or study association). Later, this change was to have important implications. But the latter part of the nineteenth century did not present the difficulty with respect to eligibility which the London society had faced. In 1902 the first assault was made upon ACS's representativeness—not on geographic, but on scientific, grounds. In that year the American Electrochemical Society was founded, arousing anxiety in the officers of ACS, who saw their society disintegrating in the face of increasing specialization.[27] In fact, the problem was a little different in this case, since a major argument advanced by protagonists of the new society was that electrochemistry is of interest to nonchemists (electrical engineers, for example) who would not join ACS and would, indeed, be ineligible. In 1906 similar arguments were advanced by biological chemists seeking independent organization. The leadership of ACS sought to adapt the organization's structure to the demands of specialization, rather as they had adapted it to those of regionalism. Thus specialist sections were formed whose officers would be *ex officio* members of the council. Meetings of the society would be held in specialist sections.[28]

But, just as adaptations of this kind had failed to prevent formation of ACS from among dissatisfied members of AAAS years earlier, so ACS was not wholly successful in preventing formation of specialist chemical bodies. In the United States—as, indeed, in England—the industrial chemists and chemical technologists proved the most militant secessionists, claiming that their writing and work was discriminated against by editors and so on, and

26   C. A. Browne and M. E. Weeks, *A History of American Chemical Society* (Washington, D.C.: A.C.S., 1952).

27   On this see Bogert, "American Chemical Societies," writing as president of ACS at a time when this was a live issue.

28   *Ibid.*

that there was little concern for their special interests. Thus the movement which developed in 1907, partly within ACS, to establish an American Institute of Chemical Engineers is important. At last, ACS was forced to recognize the rival demands of scholarship and professionalism, which had plagued its British counterpart so much earlier. In seeking to establish the Institute of Chemical Engineers, members of ACS were adamant that membership must be limited to highly qualified specialists, in contrast to ACS's abrogation of its professional function in 1902. Supporters of the new institute felt

> that a society in which these elements of maturity and experience were considered as essential would be likely to exert more influence for good in the chemical profession than to start a new society for some special chemical field with no special qualifications for membership.[29]

Thus, consideration of both the special needs of their branch of chemistry (as they seem to have regarded chemical engineering), and of their needs as professionals, appear to have motivated these dissidents. But the emphasis may have been upon the latter. In fact, ACS responded as it had on previous occasions to other groups: it established a Division of Industrial Chemists and Chemical Engineers, with its own journal. Whether because of its failure to meet the intellectual needs of the group or to address itself to the question of professionalism, the Institute of Chemical Engineers was formed in 1908.

THE PROCESSES OF CHANGE

The foregoing discussion was intended to illustrate, albeit sketchily, the sources and effects of pressures to which scientific societies are subject. In this section I propose to make a little clearer, by reference to current work of sociologists of science, what was implicit in the preceding descriptive account. We have seen that chemical societies were born of the needs of those drawn in increasing numbers to chemistry after it had attained intellectual respectability. Gradually, it became respectable to call oneself a chemist, as the discipline developed its own specific ways of investigating nature and formulating the results of its inquiries. An increasing number of scientists saw in these ways of inquiry what Ben-David and Collins, referring to a different group of scientists, called "a potential means of establishing a new intellectual identity." [30] Impetus was given the field by

29 Browne and Weeks, *American Chemical Society*, p. 88.

30 J. Ben-David and R. Collins, "Social Factors in the Origin of a New Science: The Case of Psychology," *American Sociological Review*, 31 (1966), 451–465.

creation of prestigious academic institutions concerned with chemistry, whose graduates found their way into a developing chemical industry which recognized its need of them. The generally based scientific societies, such as AAAS, fearing disintegration, attempted to accommodate themselves to demands of the new specialists by modifying their structures. But, frequently, such adaptations as they could manage proved inadequate to cater to developing needs of emergent identity.

Subsequently, in their turn, the chemical societies were forced to understand that the forces responsible for their birth were still at work as, in their turn, electrochemists, biochemists, analytical chemists, and industrial chemists (among others) sought to establish their own uniqueness. Sometimes, in these independence movements, questions of professionalism could be as basic as questions of intellectual needs, and the interests of those seeking their livelihood in chemistry were not always parallel to those whose interest was amateur—or, indeed, purely intellectual. Thus, the two pressures crucial to an understanding of the organization of chemical scientists are those deriving from the emergent needs of new specialisms and those deriving from the largely economic problems of emergent professional identity. It is these pressures, and the reactions to them, which I should like to discuss now.

Differentiation

Hagstrom has distinguished two variants on this process.[31] In the first place, he speaks of "reform specialities": subdivisions of a discipline which resent their low status within the discipline-at-large. Such low status may result from a general belief that the problems (goals) of the subdiscipline are of little consequence for the discipline as a whole. At the same time, there remain important forces retaining the reformers within their discipline. He cites as examples relativity theory within physics and logic within mathematics. "Rebellious specialities" are more seriously divided from their parent discipline since, according to Hagstrom's definition, their members "reject the claim that its goals are or should be the goals of the larger discipline" (p. 193). Under such circumstances, members may reject a low evaluation of their work, whether by editors of journals or by those responsible for allocation of rewards within the discipline. Members of deviant specialisms may then claim a greater share of the facilities and rewards available than the disciplinary status judges allot them. Such disputes may be conducted in all institutions of the discipline and may have implications for the struc-

---

31 W. Hagstrom, *The Scientific Community* (New York: Basic Books, 1965), pp. 187–195.

ture and continued integrity of all these institutions: university departments, scientific societies, grant-giving agencies. Hagstrom instances the relationship between statistics and mathematics and (although less developed) between molecular and classical biology.[32] According to this interpretation, statisticians do not feel that what they do "is" mathematics—mathematicians feel it ought to be. Once the deviants become conscious of the differences between themselves and the majority of their colleagues and come to feel that their work will not receive proper evaluation from those delegated this task by the discipline, the stage is set for strife. This may occur in all relevant organizations of the discipline. In university departments, professors working in deviant specialisms may resent the "unnecessary" burdening of their graduate students with so much "irrelevant" work. Specialists may complain that their field is denied an adequate share of journal space, or of discussion time at meetings. Under such conditions some attempt at accommodation by the organizations is to be expected, so that the discipline may retain overall control. A university may allow the deviant professor to determine the course requirements of his students; a subdepartment may even be established. In scientific societies and their journals a similar "primary adaptation" is usual. The new specialism may receive specific representation on the council of the society or on the editorial board of the journal, or a new section may be set up for it. Space in the journal may be allocated to the specialism on a quota basis (although this may be a dangerous precedent leading to relaxing standards). Sometimes this is enough and, indeed, the apparent uniqueness of the specialism may disappear. Sometimes it is not enough, and the primary adaptations may push the deviants closer together. Under these circumstances, and given certain preconditions, a new discipline will arise, phoenix-like, from conflict. Leadership is required: a few men unafraid of controversy and able to confer status on the new discipline or specialty. Status is basic, and only under these conditions will scientists

32 Mullins' account of the origins of the "informational" approach to molecular biology (the "phage group") gives no indication of any reaction of the parent discipline. This in spite of the fact that a new specialty was being carved from it, principally by physicists. (The one man who, above all, fathered the American phage group was the physicist Max Delbruck, and 41 percent of those entering the field at the postdoctoral level between 1945 and 1966 had obtained their Ph.D.s in physics.) Among social factors crucial to development of the specialty, Mullins stresses availability of funds for collaborations and summer schools, and Delbruck's "charismatic" leadership and vigrous recruiting. It seems possible from Mullins' account that if conflict with mainstream biology was avoided in the early days of phage work, this may have been a consequence of the previous status of its earliest practitioners. Delbruck, a member of the National Academy of Sciences, was able to secure publication of several papers in its *Proceedings,* while Luria (another early convert) had been editor of *Virology.* See N. C. Mullins, "The Development of a Scientific Specialty," *Minerva,* 10, No. 1 (1972), 51–82.

begin to refer to themselves as members of the new specialism. A journal of their own, granting them unimpeded access to publication, is an early requirement (if not prerequirement): "Only when a periodical is established that is devoted to a field with its own distinct goals and standards will it be possible for him to conceive of himself as a new kind of specialist." New learned societies, new university departments, new postgraduate degree courses follow.

Professionalization

What does it mean to speak of the "professionalization of science" or the "professionalization of chemistry"? "Professionalism," and the adjectives and verbs deriving from it, are terms which have been employed in a variety of ways, and some clarification of the perspective adopted is necessary. The idea of "professionalization," according to Moore, is dependent upon three crucial concepts: occupational roles, services, and professional roles.[33] An *occupational role* "denotes market-related work that has come to be sufficiently standardized so as to be recognizable by relevant lay members of society." [34] *Services* are a class of occupational roles, defined by the service concept: "Service may be defined most generally as any act of an individual so far as it contributes to the realization of the ends of other individuals." [35] Traditionally, *professional roles* have been a subclass of these service roles, dependent for uniqueness upon application of a systematic body of knowledge to problems posed by clients. Following Moore's perspective, it becomes possible to discuss professionalization, development of a new professional role, in terms of three factors. The first requirement is systematization, standardization, of knowledge in the area: growth of a consensus with respect to problems, theories, and methods. From this may flow the essential standardization of methods of utilizing the knowledge. (A prospective client must believe that the various members of a given class of professionals will offer a standard diagnosis, and prognosis, in his case.) The second antecedent is a growth in demand for the service offered.[36] Professionals cannot simply serve one another. Ben-David has pointed out that

---

33   W. E. Moore, *The Professions: Roles and Rules* (New York: Russell Sage Foundation, 1970), chap. 3.

34   *Ibid.,* p. 53.

35   Parsons, in *Encyclopedia of the Social Sciences, 13* (New York: Macmillan, 1934), 672, quoted by Moore, *Professions,* p. 53.

36   Moore suggests that high social mobility is an important factor in the growth of demand for professional services.

the establishment of a social role for scientists depended upon growth of a demand for the application of science.[37] Similarly, Moore, speaking of implications of major developments within science, points out that they can give rise to new occupational (i.e., professional) roles only when a market for any practical benefits which might follow comes into being.[38] A third requirement is that those offering the service must exercise substantial control over admission of trainees; that is to say, there must be a measure of autonomous social control over admission to the profession. In some areas this autonomy is supported by government, the professional guild being vested with not only the right, but the duty, to standardize admission. I have referred to the Pharmaceutical Society of Great Britain as having long been in this position. In the sciences, as distinct from certain professions based upon them, no such guaranteed monopolies exist; control over admission is spread through the scientific community, with corporate bodies exercising a powerful, but far from autocratic, influence. In the United Kingdom, for example, in 1969, 711 candidates entered for RIC's graduate-level professional examination; 425 were successful. At the same time, some twenty-seven hundred students obtained an equivalent qualification through graduation from a university in chemistry.

Thus the nineteenth-century professionalization of chemistry was a consequence of emergence of an intellectual consensus in the field, rapid growth in demand for chemists, and existence (in the United States, at least) of a higher education system offering appropriate training.[39]

Invariably, the need for an association involved with the economic and social status of practitioners seems to follow upon widespread diffusion of a new kind of competence within the economic system. The simultaneous growth of demand for the service in question, and of a supply of the varyingly competent, leads the most able to want their superior skills to be apparent, proven, and attested. This is achieved by creation of an exclusive

37   J. Ben-David, *The Scientist's Role in Society* (Englewood Cliffs, N.J.: Prentice-Hall, 1971), p. 34.

38   Moore, *Professions*, p. 56.

39   Between 1890 and 1910, chemists and metallurgists increased from 0.019 percent of the American labor force to 0.044 percent. (See D. M. Blank and G. J. Stigler, *The Demand and Supply of Scientific Personnel* (N.B.E.R., 1957). On flexibility of the American universities, see Ben-David, *Scientist's Role*, pp. 142–146. It is noteworthy that at Johns Hopkins courses in technical chemistry were offered in the 1870s, and in the eighties courses in electrical engineering were offered by the physics department. In 1879 research carried on in the chemistry department at Johns Hopkins led to accidental discovery of the first artificial sweetener. The German postdoctoral fellow whose happy accident it was, reputedly, caught the first train to Washington to patent the substance, later marketed under the name "saccharin." See Hugh Hawkins, *Pioneer: A History of the Johns Hopkins University 1874–1889* (Ithaca, N.Y.: Cornell University Presss, 1960), pp. 140–141, 333.

association,[40] which may come to adopt a variety of other functions (see below). Caplow has, in fact, maintained that formation of such an association is the primary stage in the process of professionalization.[41]

It is perhaps worth pointing out that the qualified chemists taken into the workshops of the nineteenth-century chemical industry carried out a wide range of functions. Research, properly speaking, was but a very small part of their work; basic research, a very much smaller part. Nevertheless, "in the less rarefied atmosphere of applied research more and more American chemists were working congenially and with gratifying results. That the operations of the industry are based on chemical reactions was so obvious that before the last quarter of the century arrived most chemical plants had a chemist or two on the premises. They had a laboratory of sorts, and although they did not often carry on research for its own sake, they analyzed raw materials and sometimes tested finished products; they tinkered with apparatus and messed about with by-products. Sometimes they went hunting for uses for new chemicals and helped good customers with their chemical troubles. All these odd jobs gradually sifted themselves out into separate functions and during these 25 years the larger companies began accumulating quite a corps of trained specialists." [42]

Implications for Organization

It is all too easy to underestimate the relevance, on the one hand, of socioeconomic factors for development of new scientific disciplines and, on the other, of intellectual factors for development of new professions. Both are stimulated by appropriate social values and forms of social organization; both require a degree of intellectual consensus. This is far from saying that the two processes of development are the same. In the first case, maturity is indicated by changes which can be limited largely to the university. A new disciplinary identity may have few implications outside the relevant area of science and external significance—though it may stimulate emergence of this identity—is no criterion of maturity. A scientific profession, on the other hand (as the term is used here), is formed more directly by market forces emanating from the world-at-large; indeed, large-scale employment may provide the major stimulus to emergence of professional identity. But this employment need not be on research, and the majority of

40 A. M. Carr-Saunders and P. A. Wilson, *The Professions* (London: Oxford University Press, 1933), pp. 298–304.

41 T. H. Caplow, *The Sociology of Work* (Minneapolis: University of Minnesota Press, 1954), p. 139.

42 W. Haynes, *American Chemical Industry—A History,* 1 (New York: Van Nostrand, 1954), p. 395.

chemists whose large-scale employment gave rise to this new occupational role were not working on "pure" research. In spite of this, processes of disciplinary differentiation and professionalization may involve a number of important similarities insofar as they affect preexisting organizations of science. Indeed, one can stimulate the other: for example, the sudden wartime need for statisticians seems to have stimulated (though not created) development of a disciplinary identity within mathematical statistics.[43] How do these processes affect scientific organization? We have seen how the initial reaction of a scientific society (like that of other scientific institutions such as university departments) is to attempt to accommodate any deviant specialism by granting concessions. Formation of a chemistry section within AAAS was an example of this primary adaptation. Later, if the rift is further widened by intellectual or social pressures, this may prove inadequate, and a new scientific society may be formed. My argument is that professionalization may exert rather similar pressures upon the institutions of science, evincing rather similar reactions.

In comparing responses of British and American scientific societies to the exigencies of emergent professionalism, the greater flexibility of American institutions (found also in universities by Ben-David)[44] is apparent. For the Chemical Society in London, primarily adaptation to demands of the emerging profession of chemistry proved impossible for legalistic reasons. A new organization, catering purely to the needs of professionals, had to be formed. The American Chemical Society did not have to face this specific difficulty until later but, when it did, it proved able to take on and discard the role of professional association as and when it chose. To some extent these differences may have been due to differences in pressures acting upon the organizations. Thus, a major objective of the Institute of Chemistry was introduction and control of appropriate training in chemistry. In the United States its sister society could rely upon the universities continuously to discharge this function,[45] partially cushioning ACS against some effects of professionalization. Nevertheless, changes in the "institutional map" of science cannot be understood purely in terms of changes in the structure of science-as-research. Market forces, growth in demand for services, may be seen to exert a powerful influence upon the "internal" institutions of science.[46]

43 Ben-David, *Scientist's Role in Society*, p. 150.
44 J. Ben-David, *Fundamental Research and the Universities* (Paris: OECD, 1968).
45 Ben-David, *Scientist's Role in Society*, pp. 142–146.
46 And the demand also for certain classes of goods, the efficient production of which requires the service of certain new classes of professions (see Moore, *Professions*, p. 164).

SOME GENERAL CHARACTERISTICS OF SCIENTIFIC AND PROFESSIONAL BODIES

Sociologists seem to have a twofold interest in scientific societies. On the one hand, they serve as indicators of change: the birth of a new society is proof of emergence of a new discipline. On the other hand, they play a crucial role in the communication system of science, largely controlling the media of communication (especially in the United States). From this latter perspective, the society is seen not as an *institution* with an internal structure, values, and interests of its own—rather, it is a nodal point within the communication system of science, and no more. Both perspectives are united, however, in regarding the scientific society as functioning solely within the social system of science, and possessing little outside relevance. Scientists join to receive publications, to attend annual conferences, and generally keep in touch: and this is their function.

Professional associations have been viewed rather differently, a consequence of the broader perspective within which professions have been studied.[47] "Organization" has been regarded by sociologists as functional for the maintenance of orderly relations between co-practitioners of a profession, and between professionals and their clients. Certainly, many of their activities may be so regarded. (A number of economists and political scientists have preferred to view them as, above all, defenders of a socially harmful monopoly. Professional ethics, from this perspective, become a means of justifying maintenance of high rates of charge. Professional associations which threaten sanctions against members who contemplate lower rates [undercutting] are defenders of economic privilege.) They are also the "voice" of the profession, both before the government and society. The scientists discover, the professionals apply the discovery: this is the usual view. And the professional deals with the outside world; the scientist, *as* scientist, speaks only to his colleagues. Thus, the function of professional associations (or the "professional" function of other associations) is much wider than the function of purely scientific associations.

The formation of an association is prerequisite for transformation of an occupation into a profession. Associations, once formed, must then seek acknowledgment as representative. This they may do in a variety of ways: by representation to employers to gain acceptance for the qualifications which they award; by receiving senior and well-known members of the occupation as members; by inducing educational institutions to organize courses leading to the association's examinations; by publications, seminars, and

47   J. Ben-David, "Professions in the Class System of Present Day Societies," *Current Sociology,* 12, No. 3 (1963–1964) for a discussion of these frameworks.

joint meetings with other groups; and by acquiring a good name for sound judgment and useful, impartial advice in representations to official committees and inquiries.[48] Millerson has suggested that their activities may be divided into "primary" and "secondary." [49]

Primary Functions	Secondary Functions
to organize	to raise professional status
to qualify	to control entry (to the
to aid the communication	extent that this is possible)
of relevant information	protection both of professionals
to register professionals	and the public
to promote professional	to act as the voice of the
ethics	profession
	social and welfare activities

Governing their relations with the outside world, and a major concern of many emergent professions, is "professional ethics." These frequently codified rules of behavior are supposedly internalized during training, and in the breach are publicly enforced by negative sanctions imposed upon errant members. It can be argued that only in cases where the client-professional relationship is one of great intimacy do such codes possess real significance. However, although this is rarely the case for professional scientists, an essential element of trust is involved even in granting a measure of work autonomy. Trust, as well as status, is an integral part of the relationship with society that the professional seeks. Hence, the emphasis upon professional codes of ethics, found in pronouncements of many technical professional bodies. American engineers, for example, have a code of ethics, drawn up by the Engineers' Council for Professional Development.[50] But could this concern with codification spring from recognition of the absence of psychological commitment resulting from the training process? And how appropriate is an ethical code emphasizing the professional's obligations to his client at the expense of those to society-at-large?

Professional associations are competitive for members as they are for status; demarcation disputes are not the exclusive prerogative of trade unions. The history of ACS, for example, includes just such a demarcation dispute with the American Medical Association over the latter's efforts to restrict chemical activity in medical laboratories to the medically qualified.[51]

48 Millerson, *Qualifying Associations,* p. 191.

49 *Ibid.*

50 H. A. Wagner, "Principles of Professional Conduct in Engineering," *Annals of the American Academy of Political Science* (January 1955), p. 46.

51 Browne and Weeks, *American Chemical Society,* pp. 228–232.

The status consciousness inherent in professional organization seems implicitly recognized by the members, and the less qualified members of professional associations participate less in their affairs. Moreover, evidence from a survey of chemists indicates that this is due partly to a feeling of inadequacy.[52] Because of this differential interest, found in the majority of voluntary associations, the leadership is rarely typical of the membership, though it is usually able to carry the membership with it on most issues.[53] Relatively few can spare the time necessary for running voluntary bodies and, clearly, there are difficulties in designing a process for the selection of leadership which gives due representation to the less successful and the newly qualified, even if this were desirable.[54] The corollary is that the leadership of professional bodies may manifest a greater-than-average interest in maintaining the status quo: management is often highly represented in the leadership of scientists' professional associations. This is not to say that policies formulated by the organizational elite are insensitive to either the views of the rank and file or those of public opinion. In time, leaders must respond to both.[55] Indeed, the influence of an organization's nominated representative at the negotiating table is often dependent upon the certainty with which he can speak for his membership. In addition, an organization dedicated to the status of its members in the world-at-large can scarcely afford to neglect the attitudes of the general public toward its policies.

Associations and Society

That both organization and policies of professional associations, including associations of scientists, are sensitive to social and economic changes in society should be clear. To recognize the extensive nature of learned societies, to see them as something more than nodal points in the communication system of science, is to understand that they are sensitive to such changes. The acquisition of professional-protective functions may be a

---

52 See A. Strauss and L. Rainwater, *The Professional Scientist* (Chicago: Aldine, 1962), pp. 172–177. Perrucci and Gerstl found that level of qualification of engineers was directly related to involvement in professional activities, so that Ph.D. > M.S. > B.S. See R. Perrucci and J. E. Gerstl, *Profession Without Community* (New York: Random House, 1969), pp. 111–116.

53 O. Garceau, *The Political Life of the American Medical Association* (Hamden, Conn.: Archon Books, 1961), chap. 3.

54 *Ibid.*

55 Thus Garceau writes, "It would be a great mistake to regard the AMA as an organization manipulated by an active minority for the purpose of perpetuating a given rigid policy" (*American Medical Association,* p. 113). "Slowly but surely, the AMA has shifted ground on such issues as medical care for the indigent. Such changes in policy have been a response both to changes in view among the membership and the pressure of outside opinion" (*American Medical Association,* p. 115).

necessary response to crisis for learned societies. Also, such responses are not evinced solely as a result of emerging professionalization of an academic discipline: they may appear as a result of changing socioeconomic circumstances.

In the early 1920s and again in the early 1930s, ACS suffered decline in membership: these were periods of economic recession in the United States, separated by a period of boom which culminated in the Crash of 1929.[56] In Britain the Chemical Society was similarly affected by the 1930–1935 depression, and its membership declined throughout the 1930s.[57] Neither society was in a position to, or claimed to be able to, secure the well-being of members. With interest in chemistry as the only criterion for membership, they were in no sense composed of professionally cohesive groups whose common interests could be defended; therefore, it is notable that the Institute of Chemistry suffered no decline in membership. Quite the reverse: in the period 1930–1939 its membership rose from 5,714 to 7,185. This was, and is, a professional association, offering the unemployed member access to an Employment Register. Thus, membership must have seemed much more than a luxury. (Curiously, however, there seems to have been little inflow of the already unemployed seeking to make use of the register.)[58] Both chemical societies reacted to crisis. Slowly but surely, ACS reassumed the professional role it had cast off in 1902. In the period 1930–1935, the bylaws were amended so that only those with "an adequate collegiate training in chemistry" as well as two years of postgraduate experience could be elected to full membership. An Employment Clearing House was established (later on a regional basis) as a result of the society's intensifying search for jobs for unemployed members. In order to assess the adequacy of college training, a Committee on Professional Training, set up in 1936, by a prodigious effort reviewed the curricula and standards of all colleges awarding degrees in chemistry. Slowly, ACS had adopted all functions of a professional body. The same course was not open to the British society, for professional activities were the prerogative of the Institute of Chemistry. A Reconstruction Committee was set up in 1930, and the major reform instituted as a result of its deliberations was introduction of local sections, with repre-

---

56 Membership fees of professional and other associations were a frequent economy at a time when salaries were often falling not only in relative terms but absolutely. See Report of the American Association of University Professors, *Depression, Recovery, and Higher Education* (AAUP, 1937), p. 136.

57 Membership figures for relevant years are as follows:

	1920	1925	1930	1935	1940
American Chemical Society	15,582	14,381	18,206	17,541	25,414
Chemical Society	3,721	4,083	3,840	3,725	3,695

58 Dennis Chapman, "The Profession of Chemistry During the Depression," *Scientific Worker* (Autumn 1939), 74–81.

sentation on the council.[59] Democratization is a standard tonic for voluntary bodies!

That scientific societies cannot be insulated from socioeconomic changes in their communities should come as no surprise. Those responsible for the leadership of such organizations, having the well-being of the institution at heart, must feel anxious to offset any tendency toward falling membership, and there is evidence that this is a likely corollary of economic depression. Though it may be analytically useful to separate "scientific" from "professional" functions, and hence associations devoted primarily to one or other of these functions, this is an artifact. Its value as a means of distinguishing associations whose relations with society are relevant from those whose relations are nonexistent or irrelevant derives from the current perspective for viewing the social system of science. This perspective is concerned with the system to the extent that it may be regarded as insulated from political, social, and economic pressures. In fact, scientific and professional functions are interdependent, and scientific organization bears witness to this fact. Students of the American depression of 1930–1935 found, for example, that the emotional disturbance resulting from financial insecurity exerted a substantial effect upon university teachers' capacity to effectively carry out their work as teachers and researchers.[60] Since, therefore, a scientist's performance within the economic sphere has implications for his performance *as* a scientist, the professional-protective function must always be a possibility for the learned society.

Today the institutions of science are called upon to respond to a set of social and economic constraints of which few scientists alive have prior experience. The continuing processes of disciplinary differentiation go on apace, and today the institutions of many newer academic disciplines, such as the social sciences, are attempting to respond to the growing extra-academic employment of members. But, in addition, we are seeing a situation of scientific un- and mal-employment which, though not on the scale of the 1930s, is more reminiscent of that time than of any time since. Also, there is disillusionment with the practical benefits which science brings, or can bring, on the part of both scientists and major segments of the community. While some may attribute it to the abuse of science by politicians, others to something inherent in science—or in rationality itself—the overall result is that the organization of science is under attack from both within and without. Throughout the world, the institutions of science are forced to respond to such pressures.

The American Chemical Society, to take one example, is now pressed vociferously to concern itself much more with its members' economic well-

59   *The Chemical Society 1841–1941*, pp. 137–138.
60   *Depression, Recovery, and Higher Education*, pp. 137–142.

being: to emphasize its professional function more, its study function less. In 1971 an ad hoc Committee on Professionalism was established under the chairmanship of the president-elect, "to consider, among other things, what the Society might do in this area, irrespective of its charter." In addition, a new Department of Professional Relations and Manpower Studies was set up. The widespread disconent of members found a focus in the person of the society's 1972 president-elect Dr. Alan C. Nixon, who intruded into "the society's largely honorific presidential election process, in which a nominating committee customarily taps two distinguished scientists for a polite contest," and was overwhelmingly elected.[61] The central issue of the "politicized" contest was unemployment. Dr. Nixon was, above all, concerned with pushing ACS into a more protective role: "I want to build a system in ACS to take care of chemists' professional needs—a system that concerns itself with the human problems of chemists as employees." [62] To further this goal, Dr. Nixon has called for establishment of a political arm to the society to engage in overt lobbying in Washington. The degree of support for these policies was demonstrated by the magnitude of Dr. Nixon's electoral victory.

But growing protectionism is not the only social pressure acting upon professional organizations. Of equal interest, and potentially of equal significance, is the attempt to redefine the *professionalism* of scientists by emphasizing the service function implicit in the claim to professional status.

Since the facilities made available for science by society (by industry, by government) are largely a function of the anticipated practical payoff, concern with application of the results of science is expedient for scientists. Today, perhaps for the first time, this may imply a concern with applicability of science to solution of the social problems of society: [63] Such concerns

> Science has received social support over the last 15 years primarily because of its role as a source of technology, but in the future it will be equally important in providing a wider intellectual base for the control and orientation of technology.

would seem enjoined upon the scientist on grounds other than that of simple self-interest, especially as a consequence of his desire to be regarded as a professional (and not *sui generis*). That scientists and engineers wish this feedback from society is clear enough from pronouncements of their organizations, and is emphasized by the findings of sociological inquiries such as

---

61 Robert Gillette, "ACS: Disgruntled Chemists Seek New Activism from an Old Society," *Science*, 173 (1971), 1218–1220.

62 *Ibid.*

63 *Science, Growth, and Society* (Brooks Report) (Paris: OECD, 1971).

that of Strauss and Rainwater.[64] Professional organizations of scientists have mainly been involved with securing newly won status and regulating members' relations with clients and employers. While few would deny that the claim to professional status, requiring the trust and esteem of society, imposes obligations upon the claimant, these duties have rarely influenced the work of scientific associations to any great extent. The obligations of professional scientists to their employers have been accorded much more importance than their obligations to society, the latter much less easily discharged. The instances of (public) dissension from the activities of governmental or industrial employers, upon the basis of which scientists or engineers might deny this indictment, are few. But of late there has been a growing awareness of these obligations, which seems to be affecting even the corporate activities of professional scientists.

To take the American Chemical Society as exemplar once more, we may note that in 1965 the society established its Committee on Chemistry and Public Affairs (made up of a mixture of academic, industrial, and governmental chemists). In 1969 this committee published its study *Cleaning Our Environment—the Chemical Basis for Action,* and it has since concerned itself with such matters as long-range energy requirements, technology assessment, and federally sponsored testing of long-term health effects of chemicals. The society has also established "Project SEED," a program of aid to the disadvantaged, involving (*inter alia*) placement of disadvantaged high school juniors in academic laboratories for summer work, and distribution of surplus journals to disadvantaged colleges.[65]

While commendable, these token activities do not discharge the responsibilities of members to society: in no way do they indicate a reordering of obligations to society and to employer, either by the profession or its corporate leaders. Among his many activities, Ralph Nader has sought to remind professionals in the United States of their duties toward society. Society may require the professional to fulfull these obligations by, for example, publicly challenging his employer when he knows the latter is acting against the public interest.[66] In Nader's view, employed professionals have rarely taken a stand against hazardous or unethical corporate behavior "as most professional codes of ethics would require," and professional associations have done little to encourage them. The fact is that professionals, especially scientists and engineers, are usually most aware of such malpractices as industrial dumping of mercury and other toxic materials, suppressed occu-

---

64 Strauss and Rainwater, *Professional Scientist,* especially chap. 9, "Imageries of Professional Status."

65 *Annual Report for 1971* (Washington: American Chemical Society, April 1972).

66 P. M. Boffey, "Nader and the Scientists: A Call for Responsibility," *Science,* 171 (1971), 549–551.

pational disease data, production of defective automobiles, and adverse and undisclosed effects of pesticides and insecticides. Unfortunately, the professional may claim, the primacy of his obligations to society is not recognized by the law, and he may find himself liable at law for damage to his employer's interests. Yet, according to Nader, "The basic status of a democratic citizen underscores the themes implicit in professional responsibility to society over that of an illegal or negligent or unjust organizational policy or behaviour." [67] Nader calls the provision of such information in the public good "whistle blowing," but there is no doubt, as he points out, that this would be commonly regarded as "ratting." Moreover, organizational norms (as well as civil law) tend to enforce the primacy of obligations to the organization over those to society. Such obligations to society might be more easily discharged if there were some clear social support for such a course of action; if "his duty to dissent" were "protected by an organization of his peers, by his professional society, and by law that requires due process and substantive justice." [68] In the absence of any lead from the professional associations—which so far have not reacted to any extent—Nader has established a Clearinghouse for Professional Responsibility to support, advise, and if necessary act upon information received. A major role of the clearinghouse is stimulation of professional associations into action as well as, of course, stimulation of individual actions. A recent editorial in the ACS publication *Chemical and Engineering News* is of interest, starting out as it does: "We'll have to admit that we were more than a little skeptical when we first read of Ralph Nader's plans to set up a Clearinghouse for Professional Responsibility." [69] The author admits that he was finally persuaded "that the concept of the clearinghouse is valid, and probably essential if scientists and other professionals are truly going to remain viable as professionals and citizens." The editorial goes on to quote with approval a remark made by Nader's associate Peter J. Petkas:

> The critical thing for a professional is his ability to dissent within the organization (without reprisal) and his ability to demand that his particular expertise not be misused.

This is a powerful but rather new interpretation of the protective function, and it remains to be seen if and when its salience today impresses itself upon the professional associations.

67 R. Nader, "Remarks at the Conference on Professional Responsibility" (Washington, D.C., January 30, 1971).

68 Quoted in *Science* (Boffey, "Nader and the Scientists.")

69 Patrick McCurdy, "Editorial," reprinted from *Chemical and Engineering News,* February 8, 1971, p. 3. Copyright (1971) by the American Chemical Society. Reprinted by permission of the copyright owner.

Interestingly enough, the American Association for the Advancement of Science, rather than any truly protective association, reacted immediately to an allegation of organizational abuse of professionalism. In December 1970 AAAS established a Committee on Scientific Freedom and Responsibility, its membership was announced in July 1971.[70] The committee had been established partly in response to requests from senators Muskie and Gravel that AAAS investigate charges of harassment of two dissident scientists (John Gofman and Arthur Tamplin) by the Atomic Energy Commission—because of their public criticism of the commission on grounds of laxity over questions of radiation safety. At that time, AAAS had no mechanism for such an investigation. The new committee has been given a mandate to

(i) study and report on the general conditions required for scientific freedom and responsibility;

(ii) develop suitable criteria and procedures for the objective and impartial study of these problems; and

(iii) recommend mechanisms to enable the association to review specific instances in which scientific freedom is alleged to have been abridged or otherwise endangered, or responsible scientific conduct is alleged to have been violated.[71]

Responses of major scientific institutions to this growing awareness of the obligations of science and scientists to *society* are few, but they are to be found. Moreover, an increasing number of individuals are attempting to stimulate them. One such is the Berkeley physicist Charles Schwartz, who for a few years has attempted to win over the American Physical Society to the idea of a Hippocratic Oath for physicists. At the 1971 meeting of the society, he also engaged in an (unsuccessful) attempt to include in the society's statement of goals a commitment to enhancement of life as well as to advancement of physics.[72] So far apathy has defeated him. It seems likely that, sooner or later, professional and scientific bodies will respond to the new social situations of science and, in particular, to the attempt to define for the first time the meaning of professionalism for the scientist, just as in the past they have responded to the growth of his professional role in society.

---

70   *Science,* 173 (1971), 129.

71   Philip M. Boffey, "AAAS: Seaborg Wins Election; Scientific Freedom Panel Created," *Science,* 170 (December 18, 1970), 1283–1285. (Copyright 1970 by the American Association for the Advancement of Science.)

72   "Briefing," *Science,* 172 (1971), 544.

THE LIMITATIONS OF PROFESSIONALISM

In their study of the American Chemical Society membership, carried out in 1960, Strauss and Rainwater found an overwhelming distaste for unionism among chemists typical of the sentiments of most professional groups: "The image of union membership is contrary to the middle-class origins of many chemists, as well as to the identifications that some men surely have made with industry and government." [73] 84 percent of members considered that the fostering of unionization of chemists and chemical engineers was a function "definitely to be avoided" by ACS.[74] Much the same was certainly true of the United Kingdom. Scientists and engineers have rarely tended to view their association with their employer as the employer-employee relationship which blue-collar workers take for granted. On the whole, they regard themselves as exchanging a service for a fee, not time for wages. In this professional association, the chemist paid to do routine analytical testing can mingle as an equal, if he wishes, with the research director of his firm: not so in a union. Union membership is often thought of as suggesting a greater degree of kinship with the line workers of the factory. To join a union is to reject the possibility of individual advancement on the basis of achievement, and to accept that only en masse can the professional and his colleagues improve their lot.

If we take commitment to such negotiation (collective bargaining) as the significant indicator of unionism,[75] we find that in a few countries it has for many years been thought compatible with professionalism. In Sweden almost all professional bodies engage in collective bargaining: the Union of Swedish Pharmacists, for example, concluded its first agreement in 1918. Today it is estimated that some 70 percent of white-collar workers in Sweden (a group which obviously includes many non- and aspirant professionals) belong to unions.[76] But in other countries the situation is very different: in the United Kingdom an equivalent figure would be between 20 and 30 percent, and in the United States probably no more than 10 percent. In neither country is there a tradition of professional unionism, and to this generalization scientists and engineers provide no exception. In the United States, the growth of large pockets of scientific and engineering manpower at the end of

73 Strauss and Rainwater, *Professional Scientist,* p. 163.

74 *Ibid.,* p. 181.

75 Postponing until chap. 5 any discussion of the limitations of this single criterion, of the relations between scientific unions and blue-collar unions, and of the kinds of unionism other than this purely "business" oriented (in Lipset's sense).

76 Arne H. Nilstein, "Sweden," in Sturmthal (ed.), *White Collar Trade Unions.*

World War II gave an impetus to unionism which was, however, soon dissipated. Several enterprise-wide unions formed at the time (e.g., in the Western Electric Division of Bell Telephone), and banded together into the Engineers and Scientists of America. This body was never able to reconcile its professional membership to an association which included technicians, and it dissolved in 1960.[77] In America, in particular, so great has been the resistance to collective bargaining that even in the Depression, during which many academics were forced to take salary cuts of 20 or 30 percent—often without consultation—few sought to involve their local branches of AAUP in an attempt to secure consultation.[78]

Given this built-in resistance, it becomes interesting to see under what conditions professionals in Britain and America abandon their traditional commitment to individualism, to status consciousness, and turn toward unionism. Second, one might ask how professional associations tend to react to such circumstances.

The following factors may be among those determining the extent of support for unionism in a profession.[79]

1. Dissatisfaction, whether deriving from conditions of work or from the prestige of the profession in the community.

2. Predisposition to militancy of the rank and file. This may be a consequence of extraprofessional statuses (social contacts in the community, social origins, sex, age, etc.), as well as of the status of the group in question within the profession as a whole and within the work organization.

3. Reactions of the authorities. Employers may inhibit unionization, either by repression and threat of sanctions (e.g., nonpromotion) or by making token concessions to satisfy many.

4. Provision of ideological leadership by a few crusading individuals.

So how may changes come about in unionization of a profession? In the first place, any specific profession must be affected by the growth of white-collar unionism generally and by acceptance of the idea that to belong to a union is not incompatible with middle-class aspirations. Even strike action is becoming legitimated as a means of protest by such groups. In a study of New York City teachers, focusing upon the growth of militancy in the 1950s

77 Everett M. Kassalow, "United States," in Sturmthal (ed.), *White Collar Trade Unions.*

78 *Depression Recovery and Higher Education* (AAUP, 1937).

79 S. Cole, *The Unionization of Teachers* (New York: Praeger, 1969), pp. 9–10.

and 1960s, Cole notes the role of deteriorating work conditions, salaries falling relative to those of other professions, and the changing sex/age composition of the profession in the city.[80] While the general tendency to radicalism of teachers relative to other professions can be explained by the generally lower-class origins of the group, growth of militancy could be attributed to the increasing proportion of young male teachers.[81] Prestige of individual groups within the profession was also important, and ideological leadership was provided by the high school teachers, an elite which had resented the introduction of a single salary scale for all teachers. Given the presence of predisposing factors such as deteriorating work conditions, relatively falling salaries, and growing youthfulness of the profession, how does the growing militancy express itself? One way, most obvious to the observer, is formation of a new organization—a militant trade union. But there is another possibility, for such an organization is unlikely to attract much professional support at the start: the professional association, to which the majority of the group belongs, may become increasingly militant.

Blackburn, in his study of bank employees in Britain, has reformulated the concept of unionization, enabling us to take account of this possibility.[82] He defines unionization by the formula "unionization = completeness × unionateness" whereby we measure extent of unionization in a field by multiplying extent of membership of each competing organization by its "trade-union-likeness." Thus, increases in unionization may result *either* from growing proportional membership, *or* from growing militancy, or both. Reviewing the growth of unionization among bank employees, Blackburn found that this resulted from growth in *membership* of the militant but initially small trade union and growth in *militancy* of the much more complete but very "gentlemanly" staff associations.[83] Moreover, in individual banks in which staff associations became rapidly militant, the union's growth was inhibited.[84] It remains to be seen to what extent these conclusions appear applicable to other professional and quasi-professional occupations; nevertheless, they seem to indicate that professional bodies can be expected to react to circumstances which might drive their memberships to militancy. Such

80   *Ibid.*

81   Radicalism in other depressed professions may be of the right: Lipset and Schwartz suggest that, when pressed to reject the status quo, "engineers opt for a movement of the radical right." S. M. Lipset and Mildred A. Schwartz, "The Politics of Professionals," in H. M. Vollmer and D. L. Mills (eds.), *Professionalization* (Englewood Cliffs, N.J.: Prentice-Hall, 1966).

82   R. M. Blackburn, *Union Character and Social Class* (London: Batsford, 1967), chap. 1.

83   *Ibid.,* chap. 5.

84   *Ibid.,* chap. 5.

reaction, in the direction of greater "unionateness," [85] is in a sense an attempt to take the wind out of the sails of any incipient, or active, union in the field. Under such circumstances, although the association may retain the trappings of professionalism, it finds itself driven to adapt its structures and functions to the membership's changing needs. It is possible, therefore, that an organization which began life as a learned society, catering to the intellectual interests of amateur and academic scientists, driven by the swelling demand for practitioners in industry to adopt various professional protective functions, now finds itself forced to modify these functions as a result of social and economic changes.

In the following chapter I shall take up this point. Is the current work and employment situation for scientists conducive to unionization? If so, how may this come about? What kinds of unions might scientists form: will they, for example, engage in political activities and associate with blue-collar unions?

85 "Unionateness" is defined by the extent to which the organization has the following characteristics: emphasis upon collective bargaining; independence of employers; acceptance of militancy and strike action; called a trade union; registration as a trade union; affiliation to the Trade Union Congress; affiliation to the Labour party. *Ibid.,* pp. 20–43.

chapter 5
Unionization and
Politicization

THE CURRENT SITUATION OF SCIENCE

In the previous chapter, I outlined factors relevant to the growth of unionization in a professional field. Among them were the following: (1) dissatisfaction, whether with work conditions, salaries, or the status of the profession in the community-at-large; (2) predisposition to militancy, as determined by extraprofessional status (such as class background, age, sex, political views); (3) provision of ideological leadership; (4) employers' reactions. Finally, it was noted that the well-known growth of white-collar unionism in general serves to legitimate unionization of a specific profession. Thus, the growth of unionism among teachers (in Britain and the United States), doctors, clerks, and other middle-class groups may help win over scientists and engineers to the view that unionism is less incompatible with middle-class identification than they had believed. Without investigating in great detail the ripeness of the scientific profession for unionization, it is possible to look briefly at the extent to which these predisposing factors seem present in North America and the United Kingdom. By this means, and by expanding a little on the notion of unionism, I hope in some measure to rationalize certain trends in scientific organization.

Dissatisfaction is widespread today in the scientific community, for

131

many reasons. In the first place, the unemployment among skilled scientific and engineering personnel is unprecedented in nearly forty years. In the peak year of 1967, the U.S. aerospace industry employed 235,000 scientists and engineers—the estimated figure for June 1972 was 147,000. Many affected by the slump in aerospace have not been absorbed elsewhere: 10,000 qualified men are said to be out of work along the periphery of Boston's Route 128 alone. An NSF survey found an overall unemployment rate of 3 percent among engineers in late 1971; the Engineering Joint Council, using a wide definition of *mal*-employment, estimated 4.7 percent.[1] The unemployed scientists and engineers are often resentful and confused, some doubting that their expensively earned skills will ever again be needed, others awaiting a change in government priorities. Interviewing a sample, a *Science* reporter found many regretful that they had ever taken up science or engineering, resentful of ex-employers and still employed ex-colleagues alike.[2] Ph.D.s and other graduates coming fresh onto the labor market soon realize that good jobs are not theirs for the asking, as they had believed. An American Chemical Society survey[3] found that only one in four 1971 chemistry graduates found jobs and that starting salaries were down 7 percent from 1970, the first time in the society's history that this had happened. Employed scientists—whether in university, government, or industrial research establishments—are finding research funds not so plentiful as formerly. University scientists may find that research grant applications are less likely to succeed: in Britain the success rate with the Science Research Council (which fulfills a function similar to the American NSF in the natural sciences) has fallen from 75 percent in 1965–1966 to 63.5 percent only four years later.[4] Growth rates in government R & D expenditure have fallen substantially. If we compare average annual growth rates in government R & D expenditures in the early 1960s with those obtaining at the end of the decade, we find that in the United States there has been a decline from 10.3 percent to —0.8 percent (at current prices), in the United Kingdom from 5.4 percent to 3.6 percent, in France from 21.0 percent to 4.3 percent, in Germany from 19.8 percent to 11.4 percent. (In terms of fixed prices, the United States, United Kingdom, and French growth rates have all become negative.)[5] At the same time, the scientist's public image has taken a battering as public interest switches from the successes of science and technology (atomic energy, space effort) to what are seen as its failures (pollution). The scientist, albeit un-

1   Data quoted in *Research Management* (March 1972), pp. 9–11.

2   Deborah Shapley, "Route 128: Jobless in a Dilemma," *Science,* 172 (1971), 1116.

3   Quoted in *Research Management*.

4   Science Research Council, *Annual Reports* (London: HMSO, 1967–1968, 1969–1970).

5   OECD (unpublished data).

willing, is cast as villain instead of hero. The extent to which science and technology have *actually* contributed to environmental deterioration is irrelevant: if there is a more real villain in every household, it does not matter. The scientist becomes scapegoat. It has become apparent that science, oversold to both public and government for too long, rarely contributes directly and immediately to the creation of either wealth or social benefit. Disillusionment is carried into the very halls of science: surely for the first time in the association's history, the presidential address to the British Association for the Advancement of Science in 1971 criticized the notion of science as inevitable harbinger of profit.[6] Finally, there are small but growing minorities gravely worried by both uses to which they feel their work is (or is likely to be) put and by restrictions placed upon their right to voice their fears.

The second relevant factor predisposing a scientific (or other professional) community to unionization involved the extraprofessional statuses of members. On average, are scientists different in age and class background from their predecessors? There is little direct evidence, but some useful indications on these points. In the first place, the rapid expansion of higher education (particularly Ph.D. output) [7] leaves little room for doubt that the scientific and technical labor force is becoming more youthful. (Indeed, 50 percent of all those qualified in science in the British labor force today are less than 35 years old.) [8] Similarly, the slow but sure "democratization" of the educational systems in many countries, allowing a growing percentage of children from working-class homes to obtain some higher education, must be producing a scientific work force enriched in professionals with working-class origins.[9] Direct data on the social composition of the scientific labor force are scarce, but the following figures indicate that an increasing percentage of American Ph.D.s come from working-class homes.[10]

6   Sir Alec Cairncross, *New Scientist,* September 2, 1971.

7   Seventy percent more Ph.D.s were awarded in pure science in the United States in 1964–1965 than six years earlier, and in the United Kingdom 105 percent more higher degrees in pure science. In Canada the output of science Ph.D.s doubled between 1961–1962 and 1966–1967; in Japan it tripled.

8   Central Statistical Office, *Qualified Manpower in Great Britain* (1966 Census of Population).

9   Few data indicate that democratization of the universities, though slow, is proceeding. We may, for example, compare the relative chances of children from lower and upper class homes for studying in universities. Some pertinent figures are as follows: United States (1958), 1:30; United Kingdom (1961), 1:8. For a few countries trend data are available: in France the relative opportunity improved from 1:84 (1959) to 1:30 (1964); in Germany it improved from 1:82 (1952) to 1:48 (1964). Moreover, science and technology, in notable distinction to law and medicine, are traditional subjects of social mobility. Data are from OECD, *Group Disparities in Educational Participation* (Paris: 1970 [mimeo], p. 65).

10   L. Harmon, *Careers of Ph.D.s* (Washington, D.C.: NAS, 1965).

TABLE 5–1
Ph.D.s in the U.S.A.

Occupational Level of Father by Year of Receipt of Ph.D.

received degree between	% with professional father		% with laboring father	
	physical science	all fields	physical science	all fields
1935–1940	33	⁚ 29	16	18
1945–1950	35	28	21	21
1955–1960	30	27	26	24

A third requirement was that there should be a measure of leadership. There must be a source of convincing argument; men capable of winning over the professional group first to the idea, then to the reality, of union membership. These might be politicians, union leaders with a penchant for public relations, or students. Students have played a substantial part in radicalization of university scientists. (I have not yet related the growth of radical feeling on political issues to the growth of unionization, but the relationship will gradually become clear.)

Fourth, employers' reactions are important, since unionization may be inhibited by either repression (and threatened sanctions) or token concessions. I do not believe this is the case today; in Britain it is the converse. The Conservative government of Edward Heath, attempting to "tidy up" industrial relations, has introduced legislation designed to foster collective negotiation. Moreover, into the "worker" (as distinct from "employer") category were lumped all manner of professional employees, including scientists (and some levels of management). When this legislation was first introduced into Parliament in January 1971, professional associations were as aghast at what they saw as forced unionization of their members [11] as were the labor unions at what seemed to them a restriction on *their* members' rights. Thus, in the apparent interests of bureaucratic tidiness, collective bargaining is receiving active encouragement in Britain.

What Sort of Union?

Unionization of a profession is a gradual process, and there is no *a priori* reason for taking it to represent total loss of concern with usual professional desiderata. Thus, although negotiation may become a group endeavor, a major feature of professional unionism may involve normal negotiation of a measure of autonomous control over the nature of the work. Indeed, a highly

11 R. Clarke, "Engineers, Unions, and the Class Struggle," *New Scientist,* (January 7, 1971), 16–18.

valued aspect of the status of professional and other high-status jobs within an organization is just this freedom from detailed supervision.[12] Thus, collective agreements may cover those aspects of the work situation which the professional (or other white-collar worker) regards as important, over which he may wish a measure of control. Salaries and salary differentials are basic, of course, but for the professional scientist so are work autonomy and (at least in theory) application of the results of his work. Whether the organizational scientist is regarded *qua* professional or *sui generis,* negotiation (either formally or informally) of a little free time for his own work is a key element in the reconciliation of many to organizational constraints. In a sense, such negotiation at the organizational level parallels negotiation between government and its scientific advisory apparatus, for the latter is often concerned with securing a degree of autonomy for the entire national scientific community (see Chapter 6). In this far from perfect world, therefore, the *negotiation* of a degree of freedom is a normative imperative for the individual scientist, and for science itself. Similar normative considerations indicate a concern with the results of science. According to the view put forward in the previous chapter, manifestation of such concern has been less than obligations to society would demand from the scientific professional. Formal discussion of the results of application of scientific research must involve the professional scientist in dialogue with government, since the latter is both the major purchaser of such application and is frequently motivated to stimulate such application.[13]

These three issues (salary questions, autonomy, and application) may be considered three alternative foci of interest or three alternative stimuli to emergent unionization. According to Blackburn's perspective, outlined in the previous chapter, unionization is measured by the "unionateness" of each protective organization active in a field, multiplied by the "completeness" of its membership. Thus, unionization may be a consequence of either rapid growth in membership of a new, highly unionate body or increasing militancy of a long-established professional association. Unionateness, it will be recalled, was measured by the score of an organization on a seven-item scale: concern with collective bargaining; independence from employers; acceptance of militancy as a possible course of action; name of the organization ("association" or "union"); registration as a trade union; affiliation to organized labor (the Trade Union Congress (TUC)—United Kingdom equivalent of AFL-CIO); and affiliation to the Labour party. In order to understand scientists' reactions to their current situation, it is useful to consider

---

12 A Sturmthal (ed.), *White Collar Trade Unions* (Urbana: University of Illinois Press, 1966), p. 388.

13 Via such agencies as the National Research Development Corporation (United Kingdom and l'Agence Nationale de Valorisation de la Recherche (France).

these items as of two kinds (even though Blackburn rejects this distinction). We may consider the latter two items (affiliation to TUC and/or the Labour party) as essentially *ideological:* membership in TUC is indicative of identification with traditional organized labor; affiliation with the Labour party indicates identification with traditional political interests of the working class. Other items are then considered as purely *instrumental.* Thus, unionateness may indicate a high ideological commitment or a high instrumental commitment (or both). Organizations formed as a consequence of pressures toward unionism may then emphasize either the ideological or the instrumental concerns (giving rise to Lipset's distinction between business and ideological unions [14] or Lockwood's between instrumental and ideological interests).[15] Thus, it is not necessarily correct to regard unionization of professionals as a consequence of developing class consciousness (as does Prandy).[16] Therefore, we may distinguish unionization as occurring by four possible routes: growth in either ideological or instrumental militancy on the part of a professional association; formation of an organization emphasizing either ideological or instrumental interests.

	Emphasis on	
	Ideological Interests	*Instrumental Interests*
Growth of New Organization	A	B
Transformation of Old Organization	D	C

Process C is relatively well known to the student of labor relations and may represent a more singleminded pursuit of vanishing status differentials, a defensive reaction to a threatened status quo:

> White-collar groups whose status is primarily based upon tradition and whose position in the plant (and consequently in society) is declining are more susceptible to unionism than groups whose status is confirmed or enhanced by the new technology and labor market balance.[17]

Of processes B and C I shall have more to say later in this chapter: they represent the two strands of present-day unionization of scientists in Britain.

14 S. M. Lipset, *Political Man* (London: Heinemann Educational Books, 1969), pp. 390–392.

15   D. Lockwood, *The Black Coated Worker* (London: George Allen & Unwin, 1958), p. 195 *et seq.*

16   K. Prandy, *Professional Employees* (London: Faber & Faber, 1965), chap. 2.

17   Sturmthal (ed.), *White Collar Trade Unions,* p. 389.

Process A may represent an outgrowth of the professional ethic. In the previous chapter I discussed current attempts to redefine the professional responsibilities of scientists, and the hitherto limited impact of such attempts upon professional associations (process D is scarcely advanced). They have, however, given rise to a number of new organizations concerned principally with these responsibilities, but which at the same time reject certain limitations implicit in a strictly professional orientation. In ideology they have advanced further to the left, in strategy they are more militant—rather than *advise* the citizen, they are anxious to *act* upon the political system. About the development of these groups I shall say more below: this is one way of regarding the new radical-scientific organizations burgeoning in the United States. In contrast to the instrumental organizations' emphasis upon salary questions, a moral concern with utilization of the results of science largely motivates these ideological organizations.

The rest of the chapter is arranged as follows: in the next section I shall discuss the growth of instrumental trade unionism among British scientists; in the section following, the political movement among American scientists; in the final section I shall attempt to draw together the two strands, considering in more general terms these extreme forms of grass roots scientific organization.

UNIONIZATION IN BRITAIN

There exist today in Britain a substantial number of unions which will accept scientists and technologists into membership. In the public sector, for example, the Institution of Professional Civil Servants (IPCS) represents the interests of all scientists and technologists (and other professional people) in government service—some 80 percent belong. IPCS takes part in salary negotiation and makes representations to official and parliamentary committees looking into (for example) the structure of the civil service or the reorganization of government R & D. According to some, it has "tended to regard its Civil Service professional objectives as the most important." [18] IPCS may be regarded as occupying an "ideological no-man's-land," being "one of those organizations which are distinctly out of line with union psychology." [19] In addition, IPCS has been under attack in the scientific journal *Nature* for failing to appreciate the need for "a more radical view of the framework within which professional scientists should be employed by

---

18 E. N. Gladden, *Civil Service Staff Relations* (London: William Hodge, 1943), p. 82.

19 L. D. White, *Whitley Councils in the British Civil Service* (Chicago: University of Chicago Press, 1933), p. 273.

the British government." [20] While IPCS has no equivalent in the private sector and no body operating in the private sector remotely matches its completeness of membership, an ideologically rather similar body for engineers, the Engineers Guild, has existed since 1938. The guild aims "to promote and maintain the unity, public usefulness, honour and interests of the engineering profession—broadly speaking, to do for professional engineers the sort of thing that the BMA does for doctors." [21] The guild was established in recognition of the limitations imposed upon the professional associations to which engineers belong by their charters. It never engaged in collective bargaining, unlike IPCS, partly because of "a certain antipathy towards trade unionism amongst a part of the membership," [22] and partly through lack of opportunity. Indeed, it did speak out against collective bargaining (in 1959) on the grounds that such procedures are "inconsistent with the professional standing of the engineer and with his proper relationship towards his employer." [23] Thus, it limited its major functions to general advice to members on salary questions and other conditions of service, to establishing standards of remuneration, and to persuading employers to require proper qualifications of their employees. Writing in 1965, Prandy referred to the scanty membership of the guild (5,000 out of an estimated potential membership of 100,000) and to its relative impotence. In part, this was a consequence of its membership, but it was also attributable to its passive role. Subsequently, the guild has been instrumental in establishment of the United Kingdom Association of Professional Engineers (UKAPE), a body with similar ideology but prepared to involve itself in collective bargaining on behalf of members. This model is an important one, on which I shall have more to say, and it is characterized in part by *reluctant* adoption of a collective bargaining function, on purely instrumental grounds. UKAPE is, in fact, rather small, although it estimates its potential membership at 250,000.

The development of bodies of this kind has not been the most *apparent* indication of growing unionism of scientists and engineers in Britain since the late 1960s. Since that time one of the fastest growing unions in the country has been Clive Jenkins' Association of Scientific, Technical, and Managerial Staffs (ASTMS). This was formed in 1968 by merger of two older unions, the Association of Scientific Workers and a technicians' union ASSET, giving a joint membership of some 85,000. Subsequently, 30,000 to 40,000 members were acquired by mergers: with a doctors' union (the Medical Practitioners Union) and insurance staff associations. Deliberately,

20   "Compromise Without Solution," *Nature,* 232 (1971), 593.

21   Prandy, *Professional Employees,* p. 75.

22   *Ibid.,* p. 79.

23   *Ibid.,* p. 80, quoting Engineers Guild, "Collective Bargaining" (1959).

its dynamic extrovert general secretary "turned his attention to insurance and bank staffs, publishers, university lecturers and technicians and doctors as well as middle to senior managers." [24] As this commentator goes on to say, "ASTMS is now beginning to look more like a cadre organization—the 'white-collar TUC's' of Sweden, Germany and France—and less like a single union." [25] Undoubtedly, a growing percentage of scientists are among the union's 220,000 members. The association has competed militantly for the right to represent substantial groups of scientific workers, and many large employers have been impressed with the benefit of having all their white-collar workers represented by a single negotiating body. The union has signed an agreement with the French Confédération General de Travail (CGT) designed to secure international action, when necessary, against multinational companies—specifically, in the fields of automobile manufacture, chemicals, petroleum, and aeronautics.[26]

The much publicized growth of this militant version of traditional unionism among professional groups has exerted a marked effect upon the professional establishment. In the first place, together with more fundamental causes, it has stimulated a measure of increased unionateness among the professional bodies. In the second place, it has served to stimulate manifestation of substantial disapproval of the appropriateness of traditional working-class unionism among and for professionals.

Many professional bodies have for some time exercised some marginally union-like functions. Many carry out surveys of members' earnings: this practice was instituted by the Royal Institute of Chemistry (RIC) in 1930. RIC has also drawn up a scale of "appropriate" salaries. Every year the British Association of Chemists (a small registered trade union formed in 1917) also does a "Salary Scales Assessment," a computation related to cost of living, profitability of industry, and qualifications of members. These salaries are intended as minima. But activities of the professional associations are restricted by their charters: since both managers and employees are represented, they are not allowed to take part in collective bargaining. Quite recently a working party of the Council of Scientific and Technical Institutions (CSTI), to which all belong, was established with a mandate to determine whether, in view of the prevailing economic situation, scientists and technologists could be accommodated within the existing union structure. The feeling was that they could not, and concern was exacerbated by the new industrial relations legislation introduced in 1971 by the Conservative government. Though the unions (such as ASTMS) claimed to

24 John Torode, "Brother White Collar," *New Statesman,* September 3, 1971, pp. 292–293. (Reprinted by permission.)

25 *Ibid.*

26 "French and United Kingdom Unions in Pact," *Observer,* March 14, 1971.

look after scientists' interests, CSTI doubted this, on the grounds that scientists' interests would be subordinated to those of the numerically larger groups of technical-level staff. In March 1971 the RIC circulated a document among its members which stated:

> Staff unionisation in British Industry is spreading rapidly and many professional people are beginning to be affected, including chemists and other scientists. This trend will almost certainly continue and the pace must be expected to accelerate. The Council of the R.I.C. believes that the general majority of chemists do not oppose collective bargaining in principle, although some may have reservations about its effectiveness at professional level. They would, however, strongly resist attempts to force them to join, or be represented by, unions that are unlikely to pay due regard to their interests or whose tactics are not in keeping with professional ethics.
>
> In common with other professional institutions, the R.I.C. . . . is also actively collaborating with other science institutes in forming a protective body for scientists and technologists which, if circumstances require, will be able to register as a trade union. It will be called The Guild of Professional Scientists and Technologists.[27]

The views now expressed by RIC are in contrast to those once put forward by the Engineers Guild on collective bargaining. The Industrial Relations Act has provided the impetus for organization of a trade union for scientists and technologists irrespective of area or sector of employment. There would appear to be a widespread feeling among leaders of professional associations that, if professional men are likely to obtain significant advantages from belonging to a registered union, that union should be under their aegis or control.

In spite of general professional dislike of the notion, of the associations, and of unionism, many professionals have long been involved in traditional trade union activities. Since 1917 there has been a trade union for chemists (the British Association of Chemists), while 40 percent of the membership of the Royal Institute of Chemistry works in areas of employment in which negotiating machinery or collective bargaining arrangements have long been in operation. The scientists belong to staff associations, protective organizations, or unions; most are officially recognized in their respective spheres of interest. RIC members have joined such bodies voluntarily to benefit from their functions as bargaining agents: membership has rarely, if ever, seemed to conflict with chemists' professional ethics. "Professional unions" have tried to avoid the ideological commitments of orthodox unions, offering their

27 *The Industrial Relations Bill—Its Probable Consequences for Chemists and the R.I.C.* (London: Royal Institute of Chemistry, 1971). The title "Guild" has been replaced by "Association."

status-conscious membership all the advantages with none of the disadvantages of belonging to a union. To join a manual trade union, or any welcoming into membership low-level scientific technicians, has been widely regarded as against the interests of graduate-level scientists and technologists, since such a union may be committed to reduction not maintenance of salary differentials. The Association of Professional Scientists and Technologists (APST) will press for maintenance of differentials favorable to its members. Those responsible for creation of the new organization, covering all sectors of employment, believe they will answer a real need. As indicated, the association aims to represent the interests of a wide range of professionals (including chemists, physicists, metallurgists, mathematicians) in negotiation with a wide range of employers, and on all aspects of work conditions.

Though this diversity offers the new body the advantage of size, its organizers appreciate that weakness may also spring from diversity. Given such a variety of qualifications, capabilities, and concerns, how can the union satisfy each individually while preserving the essential united front before employers? Take, for example, the question of the establishment of wage norms and minima—an important part of trade union activity. Such procedures depend upon job evaluation. When members carry out rather similar tasks, as on an assembly line, neither evaluation nor establishment of norms is too difficult. But this is very different from consideration of the complex and multifarious tasks which chemists alone may be engaged upon: ranging from basic research, through development, product improvement, trouble-shooting, to technical salesmanship. Job evaluation becomes extremely difficult: establishment of agreed norms of payment even more so. The heterogeneity of the membership seems likely to produce pressure group activities among different groups, designed to secure differentials within the membership. On the face of it, this seems an inherently unstable situation for a trade union. It is accepted that in negotiation factors extrinsic to the work will affect wage assessments: region of employment, environment, facilities at workplace, and so on.

The professional bodies have been *pushed* not altogether happily into an uncertain, ambiguous situation which they must now attempt to rationalize: "The Council has always taken the view that there can be no question of regarding collective bargaining or membership of a union as being inconsistent, in principle, with professional ethics. On the other hand, it has also firmly maintained that the indispensable features of professional conduct include the duty to act with dignity and restraint, to honour obligations and to give due regard to the public interest." [28] This ambiguity is reflected both

28 *Ibid.*

in a general anxiety to model the new body on the British Medical Association and in the view taken of links with the Labour party and the Trades Union Congress.

With the exception of ASTMS (in which the political levy making up the union's contribution to the Labour party is voluntary), none of the organizations to which I have referred has any formal relationship to the Labour party or even (with the exception of ASTMS) to TUC. The chemists' trade union, the British Association of Chemists (BAC), has no such links: "In 1920 the Association was registered as a trade union, and as such it is empowered to conduct negotiations with employers on behalf of members. It is, however, completely nonpolitical and it is not affiliated to TUC. BAC has always maintained that membership of a trade union for chemists must take due regard of the professional obligations and dignity of the members." In view of APST's dislike of orthodox unionism, at least in its own field, it is not likely that it will make the symbolic political gesture which affiliation with the Labour party (or any other party) would represent. Whether or not it might at any stage affiliate with TUC is less easy to predict on *a priori* grounds: many white-collar unions have found it difficult to decide the precise nature of their relationships with the congress. IPCS, for example, representing scientists and engineers in the public service, has made tentative inquiries about affiliation, while the Association of University Teachers sends observers. It has been suggested that APST would affiliate if TUC were more prepared to act in concert with the government and the employers (Confederation of British Industry). The general uncertainty with which white-collar unions view affiliation is shown clearly enough by the local government officers' union NALGO which joined after three attempts. The membership was very hesitant, and the association admitted that it might feel itself forced to disaffiliate at some stage. These organizations find it difficult to balance the practical advantages of membership (especially economic support) against the symbolic identification with manual labor which membership entails.

APST's immediate aims are strictly practical: to seek recognition as a proper negotiating agent under the Industrial Relations Act and to offer the advantages (but none of the disadvantages) of union protection. The various forms of union militancy, though disliked by professionals, have not been wholly ruled out. BAC, while opposing strikes, has not ruled them out, and it includes a kind of scientific "work to rule" in its armory. More usual is the blacklisting of employers, with whom members are discouraged from seeking jobs. APST's array of weapons is likely to be similar: when membership is large, and a defense fund established, the association may be in position to withdraw members from particular sections of industry or firms, offering them interim financial support.

When APST is established, the professional associations will retain their traditional functions. RIC observed as long ago as October 1970 that it would shortly undertake increased activity on behalf of the status, economic interest, and general welfare of members, and that these responsibilities would be fulfilled by either the institute acting alone or

> if desirable in collaboration with other bodies such as staff associations, trade unions, and employers associations.[29]

The possible need for a new organization (APST) was then foreseen. In brief:

> Its (APST's) sole purpose will be to undertake those protective activities in the field of collective bargaining that the RIC and the other scientific institutes may not be in a position to carry out effectively because of their wider public responsibilities.[30]

Its potential membership is estimated at 35,000, with the likely membership estimated at 10,000 after one year, rising to 20,000 by the second, and to 25,000 by 1975. Virtually all will be qualified to degree level.

Relations between the "status" unions being formed in this way by the professional associations and the white-collar unions, such as ASTMS, recruiting in the same occupational area, are not marked by a simple spirit of healthy competition. Ideological differences run deep. An official of a new professional union, UKAPE, has written of traditional unions that they

> are committed to the principles of the closed shop and strict demarcation, on the assumption of one union to each man, limitation of individual freedom in the name of discipline, percentage rises irrespective of merit, and progressive elimination of differentials.

This philosophy does not appeal

> to the professional class, who take their economic survival for granted and are far more interested in their status in society. They are interested in salary in terms of the status it brings rather than in terms of beer and cigarettes. This means that the way they get increases is as important to them as the increases they get. Is the increase a recognition of good work or has it merely been extorted? [31]

29   *The Future of the Institute as a Professional Body* (London: RIC, October 1970).

30   *Industrial Relations Bill—Its Probable Consequences* (London: RIC, March 1971).

31   Clarke, "Engineers, Unions, and the Class Struggle." (This extract first appeared in *New Scientist,* the weekly review of science and technology, 128 Long Acre, London WC2, and appears with the publisher's permission.)

These differences in viewpoint have been brought into the open by the recent case of Mr. J. W. Hill, a contract engineer employed by an engineering firm based in Newcastle. A few years ago this firm recommended that all its staff join a union, ASTMS and DATA (Draughtsmen's and Allied Technicians' Association) being recognized for negotiating purposes. But DATA wanted sole rights, and its members engaged in strike action to secure them. In May 1970 the company agreed that all its technical staff should join DATA, and that this should be a condition of employment. Mr. Hill and thirty-eight colleagues refused to join the union, being members of UKAPE and believing that membership in a union prepared to take strike action was incompatible with their obligations as professional engineers. Subsequently, the dispute was carried to the courts, generating a good deal of rhetoric.[32] The president of the Institute of Mechanical Engineers has referred to the "unfair position" in which the engineers had been placed.[33] The chairman of the Engineers Guild has affirmed his approval of the engineers, preferring to "face dismissal rather than undertake to transfer from a union which recognizes their integrity to one which irremediably compromises it." He takes exception to the political commitment implied by membership in a traditional union (which may not be accurately reflected in the contribution of the union to Labour party funds): [34] "Their politics and methods are those of extremism and aggressiveness . . . they seek to threaten the livelihoods of fellow workers whose only offence is unwillingness to subscribe to their industrial and political aims (which go far beyond the normal purposes of union representation)." Representatives of the orthodox unions have responded by referring to the "snobbery" inherent in these views, and to what they see as the practical similarity of the relationships between employer and engineer and employer and blue-collar worker. This similarity suggests, or ought to suggest, an identity of interest—although, of course, engineers feel additional concern over issues such as salary scales, career structures, and facilities for obtaining additional qualifications [35] and are more likely to join the managerial class eventually. But on the same day the correspondence columns of *The Times* also carried a letter intimating that "it is not against the employer that the professional employee feels the need to be 'organized.' " [36]

The disenchantment of British scientists appears to have given rise to a unionization movement proceeding by two separate mechanisms. On the

32 "Law Report," *The Times,* August 23, 1971.
33 Letter from R. L. Lickey, *The Times,* August 27, 1971.
34 Letter from F. A. Sharman, *The Times,* September 5, 1971.
35 Letter from R. Miller (ASTMS), *The Times,* September 4, 1971.
36 Letter from B. Bransbury, *The Times,* September 4, 1971.

one hand, one may note the expansion of traditionally based white-collar unions (with their traditional commitment to the Labour party) as a consequence of militant recruiting. On the other hand, the gradual, reluctant adoption of certain union-type functions by professional associations (albeit, for legal reasons, via formation of new bodies) is also apparent. The ideological differences between the two strands are clear enough, indicative of identification with the interests of employer and workers. Unionization may or may not represent the development of class consciousness among professionals. Ideological concerns may be minimal and, where present, they *may* spring from a radicalism of the right.

POLITICIZATION IN THE UNITED STATES

Discontent over the finance, control, status, and especially *application,* of science has given rise to a more specifically ideological movement in the United States, operating essentially on the political plane. Again, however, various strands to the movement must be distinguished. The first major manifestation of grass roots political concern within the scientific community of America followed the development of the atomic bomb and the subsequent arrangements made for control of atomic energy. Physicists and chemists, specifically those working in the laboratories of the Manhattan Project in which the atomic weapons research had been concentrated, became increasingly worried over the implications of what they had done. Organization of the scientists was precipitated by the nature of the postwar legislation introduced for the control of atomic energy,[37] allowing for continued military control with its attendant secrecy. In an attempt to defeat this legislation (the May-Johnson Bill), which seemed certain to imply that atomic energy would remain forever and above all the basis of a weapon, the scientists began to mobilize. But it was not only the apparent use of their work which scientists found repugnant; they were also alarmed at the functioning of the decision-making process.[38] The atomic scientists were unhappy that their "spokesman," the scientific advisors in positions of power (e.g., Bush, the Comptons, Fermi, Lawrence, Oppenheimer), appeared to be approving a policy which they overwhelmingly disliked. Out of these worries came the Federation of Atomic Scientists. Founded in December 1945 from various independent laboratory groups, its objectives were indicative of the nature of its concerns: [39]

37 Alice Kimball Smith, *A Peril and a Hope* (Chicago: University of Chicago Press, 1965), chap. 3.

38 *Ibid.,* p. 136.

39 Objectives quoted, *ibid.,* pp. 236–237.

1. In the particular field of atomic energy, to urge that the United States help initiate and perpetuate an effective and workable system of world control based upon full co-operation among all nations.

2. In consideration of the broad responsibility of scientists today, to study the implications of any scientific developments which may involve hazards to enduring peace and the safety of mankind.

3. To counter misinformation with scientific fact and, especially, to disseminate those facts necessary for intelligent conclusions concerning the social implications of new knowledge in science.

4. To safeguard the spirit of free inquiry and free interchange of information without which science cannot flourish.

5. To promote those public policies which will secure the benefits of science to the general welfare.

6. To strengthen the international co-operation traditional among scientists and to extend its spirit to a wider field.
   We shall endeavour to keep our members informed on legislative proposals and political developments which affect the realization of our aims, and to co-operate with other organizations in the achievement of these aims.

After going into a substantial decline, this federation, under a new name, flourishes again today, preoccupied with new issues. But now it is but one of a number of organizations concerned with such matters, and among these organizations significant differences exist. One basis for distinguishing between them is in terms of the political form which their ideological concern takes.

In his essay "Politics as a Vocation," Max Weber made the distinction between the "ethic of responsibility" and the "ethic of ultimate ends": [40] under the former, one is concerned with people's weaknesses and one's responsibility for them; under the latter, with the morality of one's objectives only. From this distinction may be deduced a framework for analysis of political behavior which serves our present purpose well. We may distinguish between "goal-oriented" and "expressive" politics: the first concerned with achievement of specific political objectives, the second with satisfactions to be derived from political activity in itself. *Goal-oriented politics,* then, may be regarded as the pursuit of power and influence, often over specific issues of policy; *expressive politics* often involves sacrifice of power to principle, and may be regarded as the sheerest manifestation of principle. Parkin attempted to apply a distinction of this kind to analysis of the British Campaign for Nuclear Disarmament (CND) in the mid-1960s.[41] He concluded

40 H. H. Gerth and C. W. Mills (eds.), *From Max Weber* (London: Routledge & Kegan Paul, 1948), pp. 120–121.

41 Frank Parkin, *Middle Class Radicalism* (Manchester: Manchester University Press, 1968), chap. 3.

that, with its emphasis upon essentially "symbolic" marches, with its tacit recognition of its political ineffectiveness, CND had to be regarded as an institution committed to the expressive form of politics. The social position which the majority of CND supporters occupied was preeminently middle class, and Parkin sought to explain their radicalism in terms of their rejection of middle-class values:

> Those who subscribe to deviant values, particularly of a religious or political kind, will have to seek outlets for their expression and reaffirmation since they are not firmly institutionalized in the social system in the way that dominant values are. . . . The marches, demonstrations, vigils and so forth which characterize expressive politics provide a means of re-inforcing values which are not securely integrated in the social structure in the sense that they lack the support and legitimating effect of major institutions.

Thus, it may be hypothesized that rejection of the values of science by society (whether implicitly or explicitly) or an apparent "selling out" by a major part of the scientific community could lead to a phenomenon along these lines somewhere within that community. A final characteristic of this kind of political activism, according to Parkin, is that a single protest movement is likely to constitute symbolization of deviant views on a whole range of issues, and is rarely likely to restrict its concern to a single issue.

I believe that this distinction provides a useful means of understanding the variety of political protest movements springing up on the periphery of American science and, just as important, the basis of their disagreement. While some of these organizations are committed, above all, to modifying national policy on a limited range of issues, others are dedicated principally to symbolic demonstration of the ethical failings of contemporary science.

I propose now to discuss five of the better known political groups developing from within the scientific community in the United States, together representing the new "critical movement" in American science.[42] The organizations are the Federation of American Scientists (FAS), the Council for a Livable World (CLW), March 4th/Union of Concerned Scientists (UCS), the Society for Social Responsibility in Science (SSRS), and Scientists and Engineers for Social and Political Action (SESPA). Specifically environmental groups are excluded, as are all those in which scientists operate essentially as advisors to citizens: the scientists' information movement, which has something in common with these organizations, is described elsewhere in this book.[43] The first two organizations (FAS, CLW) are very definitely "goal oriented" in terms of the distinction made earlier; SESPA

---

42 Sometimes called the "New Critics"; e.g., see M. L. Perl, "The 'New Critics' in American Science," *New Scientist,* April 9, 1970, p. 63.

43 See chap. 7.

(I shall suggest) is essentially an "expressive" organization. I will discuss each organization in turn, attempting to keep to some sort of standard outline.

(1) The Federation of American Scientists

ORIGINS.   The Federation of American Scientists (FAS) developed out of the Federation of Atomic Scientists, whose origins have been so well described by Alice Kimball Smith.[44] We have alluded to its founding in opposition to the proposed system of control of atomic energy put forward in 1945 (in the May-Johnson Bill), essentially among the atomic scientists so profoundly shocked by Hiroshima and Nagasaki. The federation sought to educate both Congress and the American people to its view of the need for international control of the atom. To this end it cultivated relations with the press, labor, and religious organizations which might be sympathetic— even with Hollywood (briefly!). At the same time, a Washington office was opened in recognition of the importance of direct lobbying of congressmen. There seems to have been some difference of opinion over relative importance of these two sorts of activities,[45] and this had implications for the organization's status since as an educational body it could claim tax-exempt status, but this precluded lobbying. It was largely as a consequence of the special relationships established with friendly staff members of crucial congressional committees that FAS was able to help defeat the May-Johnson Bill. For one must remember that the leaders of the organization were not the experienced statesmen of science.

To a substantial degree, success of FAS derived from the American system of government: more specifically, from the power and independence of Congress. There are few causes which cannot find a friend on Capitol Hill and many may find a friendly committee, with a friendly expert staff. Given this, plus the jealously guarded power and independence of the American legislature, lobbying groups have an avenue to effective influence denied equivalent organizations in many countries. Thus, while in the House the May-Johnson Bill came before the Military Affairs Committee, in the Senate a jurisdictional dispute led to establishment of a new committee specifically to conduct atomic energy hearings. Its chairman was Senator MacMahon, who would prove FAS's closest ally in the Senate, and the staff of his committee became the federation's most important tactical advisors in matters of politics.

Subsequently, FAS entered into a decline. With the temporary exception of the tremendous attacks upon Oppenheimer and Allen V. Astin (then

44   Smith, *A Peril and a Hope.*

45   *Ibid.,* p. 292.

director of the Bureau of Standards), few issues united the membership as had the attempt to secure civilian control of atomic energy. FAS found itself divided, for example, over the morality of U.S. importation of ex-Nazi scientists,[46] and over the proper policy to adopt when truly international control of atomic secrets became a patent impossibility. On this latter issue, Smith observes clear lines of demarcation between the elder scientists, regarding any rapprochement with the USSR as an impossibility, and the younger ones, hoping to induce a more cooperative mood among the Russians.[47] Issues seemed to become more political, and it was widely believed that for scientists to take sides would endanger the political prestige so recently won. The prestige, after all, derived from expertise: they had been able to establish themselves as indisputable experts on atomic science —as expert as anyone the "opposition" could put forward. They could not speak with authority or claim expertise on many new areas of contention. Thus, as the "Atomic Scientists" became the "American Scientists," as the basis of eligibility for membership was widened, as the range of interests grew, the organization's influence began to decline. Soon this was reflected in membership statistics: from a high of three thousand, it fell to only about one thousand in 1949. FAS lapsed into near invisibility, with only occasional, transient dashes into the limelight.

FAS TODAY. But the federation has recently begun to bestir itself: launching a membership drive, reopening an office on Capitol Hill. In 1969 the decision to appoint a full-time lobbyist-director was made by the council: this was regarded as crucial to any attempt at revival. To this post was appointed Jeremy J. Stone, thirty-five-year-old mathematician-turned-arms-specialist. At the same time, the key decision to stick to the same kind of "elitist" lobbying so successful years before (but now from a hopefully expanded membership base), in preference to any New Left orientation, was implemented. So, once more, FAS started to solicit the membership of scientists "to help the Congress prevent the misuse of science; to prevent such boondoggles as ABM and SST; to promote industrial conversion to peaceful purposes; to organize student interest in science-and-society problem solving; and to construct a voice of science that can really be heard." [48] A publicity leaflet quotes Jerome Wiesner (President Kennedy's science advisor):

> There is no other group that so truly represents the conscience of the American scientist and no other group that has worked so hard and long

46 *Ibid.*, pp. 339–341.
47 *Ibid.*, p. 506.
48 Publicity leaflet, Federation of American Scientists, 1971.

to understand and explain the complex, dynamic expanding scientific activities which so thoroughly dominate and perplex the lives of the people of our time.[49]

And Senator Philip A. Hart:

> The type of specialized assistance that your organization provides the members of the Senate, in such timely and useful manner, is enormously helpful to us as we tackle these very complex problems.[50]

These quotations give an idea of the self-image of the federation: liberal, respectable, "The Voice of Science on Capitol Hill." So liberal and respectable, indeed, that many younger, more radical scientists find it far too friendly with the Pentagon, far too happy to operate within the corridors of power.[51] Unlike them, it concerns itself far more with specifics of policy than with rhetoric or with the ethical basis of scientific behavior.

RELATIONSHIPS WITH THE SCIENTIFIC COMMUNITY.   The federation is small but growing: from about fifteen hundred in July 1970 to about twenty-three hundred in July 1971. In spite of its grass roots base, the leadership of the federation is closely related to the liberal Establishment of American science: its sponsors and council include many of the most famous names in American science. Sponsors include Nobel laureates Hans Bethe, Owen Chamberlain, and Harold Urey; ex-presidential science advisors Kistiakowsky and Wiesner; and economist J. K. Galbraith. Chairman is Herbert F. York, one-time director of the Livermore Laboratory, ex-chancellor of the University of California at San Diego, first director of Defense Research and Engineering (in the Department of Defense under Eisenhower), and ex-vice-chairman of the President's Science Advisory Council (PSAC). Vice-chairman is Marvin L. Goldberger, head of the Princeton physics department and one-time member of PSAC. Council members include Holton (Harvard), Luria, Morrison, and Weisskopf (MIT), and social scientists such as Halperin (Brookings), Rathjens (MIT), and Capron (Harvard). The organization is run by the same men, many of them physicists, who until recently governed science, from NAS, PSAC, and secret defense policy committees. Goldberger, for example, once chaired the strategic weapons panel of PSAC. Presumably, his "conversion" is typical:

49   *Ibid.*

50   *Ibid.*

51   Deborah Shapley, "FAS: Reviving Lobby Battles ABM," *Science,* 171 (1971), 1224–1227.

You give excellent advice but you are always 180 degrees out of phase, and you become bitter. And so you are forced to the conclusion that it can't be done from the inside and one simply has to work from the outside.[52]

But in moving "out into the cold," scientists such as Goldberger and York carry with them an unrivaled knowledge of the mechanics of decision making, of where and how to apply pressure, crucial to the kind of operation mounted by FAS.

ACTIVITIES OF FAS. The critical organizations of science may be characterized in terms of membership and leadership, activities, and issues concerning them. In discussing organizational activities, it is of interest to try to ascertain the balance between "instrumental" and "expressive" or "symbolic" modes of action. For example, congressional testimony, or the support of congressional candidates, is almost certainly instrumental (the attempt to influence specific legislation), in contrast to "research strikes" or "marches" which have high symbolic importance.

Congressional testimony is a major weapon in FAS armory. A good deal of importance is attached to provision of information designed to enable friendly congressmen to bolster their cases. For example, "FAS intends to seek tax-exempt funds with which to pay per diem fees to academic specialists who would not otherwise be able to provide their services to interested Congressmen." [53] On occasion, active lobbying may support these informational activities. During the 1970 congressional debate on the ABM, "the Federation devoted virtually all of its time to the ABM. We presented approximately 70 Senators—all who might vote with us—summaries of the testimony on ABM before the Armed Services and Foreign Relations Committees." [54] Use can be made of the prestige membership of FAS:

It became evident that the Senate had lost sight of the fact that Safeguard was worthless even if it worked perfectly. To drive this point home, three sentences of testimony by W. K. H. Panofsky were put together and agreement to these sentences was secured from Jerome B. Wiesner, Herbert F. York, Herbert Scoville, Jr., Marvin L. Goldberger, and (later) Donald F. Hornig.[55]

At the same time, attempts may be made to mobilize the mass of the scientific community on such issues:

52 Andrew Hamilton, "March 4th Revisited Amid Political Turmoil," *Science*, 167 (1970), 1475.
53 *Ibid.*
54 "Report on the ABM Debate," FAS *Newsletter*, 23, No. 7 (1970), 2–3.
55 *Ibid.*

Earlier in the debate, the Federation mailed to each of the first 1000 (thousand) physics departments a copy of [the statement on Safeguard with a chart prepared by Panofsky] urging scientists to send telegrams.[56]

Statements are released to the press on major issues, stimulating discussion in the press. Symposia are arranged, for example, on the "Ph.D. surplus" in conjunction with the 1970 meeting of AAAS. Other grass roots activities are in process of organization. Among these is TACTIC (Technical Advisory Committees to Influence Congress). "In each Congressional district a Technical Advisory Committee to Influence Congress (TACTIC), consisting of about half a dozen scientists and engineers will be recruited. With the (national) Issues Committees providing position papers and suggesting other references, each TACTIC will contact its Congressman and advise him on issues of concern to the scientific community . . . the TACTIC group will urge him to take a personal interest in these issues." [57] So far about five hundred scientists in 225 congressional districts have enlisted to form local TACTIC groups, in the hope of generating effective local pressure.[58]

ISSUES REPRESENTATIVE OF THE CONCERNS OF FAS.   Three issues on which the federation has recently demonstrated substantial concern are the strategic arms race, SST, and unemployment of American scientists. Its attitude on these issues and the style in which it is expressed may be deduced from the federation's monthly *Newsletter*.

On October 26, 1970, a statement was issued rebutting the claim of the American Security Council (ASC), a private body largely financed by industry, that the United States was lagging behind the USSR in the strategic arms race. The statement was prepared by an FAS subcommittee chaired by Dr. Herbert Scoville, Jr. (former deputy director of the CIA and assistant director of the Arms Control and Disarmament Agency). The essence of the critique of ASC's assertions was to be found in the different measure of position in the arms race employed:

> That U.S. security is dependent on total megatonnage is basic to the American Security Council fears. This outmoded concept was discarded by U.S. military leaders many years ago. Numbers of warheads and bombs and their invulnerability, penetrability, and accuracy are much more significant criteria.[59]

---

56   *Ibid.*

57   *Ibid.,* p. 1.

58   Shapley, "FAS: Reviving Lobby Battles ABM."

59   FAS *Newsletter,* 23, No. 10 (1970), 3.

At a press conference enunciating the statement, carefully documented statistics of deployment on this basis were presented. An important aspect of recent policy has been introduction of multiple independently targetable warheads (MIRVs), increasing the number of warheads but decreasing the total tonnage.[60]

> As then Deputy Secretary of Defense Paul N. Nitz testified in 1967, the introduction of these multiple warheads provides "much more effective payloads" by "every relevant criterion" of military effectiveness despite their overall lower megatonnage. . . . Limitations on MIRVs are a prime objective of arms control, and the Federation of American Scientists has supported this goal.

The rebuttal concludes:

> We have attached an authoritative table of U.S. and Soviet strategic forces that is based on Defense Department officially released information which may be used in place of the Operation Alert chart.

The statement is in measured language, deals with a specific issue, and is based upon clearly presented concepts and data taken from official sources. The same careful tone is found in later statements on the same issue:

> We therefore urge the Administration to make vigorous efforts to achieve a comprehensive freeze agreement that precludes the wholesale replacement of existing weapons with more advanced versions. Such an agreement would be consonant with the Administration program of "negotiation not confrontation." The U.S. Senate has already expressed widespread bi-partisan support for an agreement of this kind.[61]

Perhaps because it is attempting merely to affect specific policies within a status quo which is not rejected out of hand, FAS attitudes are likely to receive a degree of support among the uncommitted, given adequate publicity. In a recent report issued in May 1971, and in testimony before Congress, the federation has been working hard to discredit DoD predictions of Soviet military-technological superiority (linked to requests for increased departmental funds). *Science* finds that the scholarly, reasoned approach "has achieved surprising success in its challenge to the Pentagon. Several influential members of Congress have listened to their arguments attentively."[62]

---

60 *Ibid.*

61 *Ibid.*, p. 3.

62 Robert J. Bazell, "Arms Race: Scientists Question Threat from Soviet Military R & D," *Science*, 173 (1971), 707–709.

A similar approach typifies FAS's approach to other issues, such as the Supersonic Transport (SST). In spite of widespread lobbying in its favor, SST was'voted down by Congress at the beginning of 1971. Among those protesting against the project was the FAS. A federation statement released in September 1970 was aimed at setting forth objections to SST.[63]

> SST proponents estimate that 10 percent of our population will be flying internationally in the latter part of the century. Only a well-to-do fraction of these will use the expensive SST to save only a few hours in most cases and a half-day in others. Meanwhile the Government is planning to spend almost three times as much on the development of the SST *alone* as it is planning to spend over the next *12 years* for research and development on new forms of mass transit.
>
> Further the SST prototype program is a poor "business" investment for public monies. Since SST is not a high priority project, the same reasons that the Department of Transportation explained were an "insurmountable hurdle" to attracting private funding should preclude Government financing—the long "dry period" before profits, the "considerable technical risk" and the "amount of profit which would finally accrue" . . . the Government return under this contract, even if all goes well, is conceded by the Department of Transportation to be only "a little over 4 percent."
>
> The SST is an environmental hazard. No one can doubt that Government rules on noise and sonic booms will be bent, if necessary, to keep the finished SST aircraft economically viable. . . . The dangers of pollution of the upper atmosphere, even if in fact quite serious, could not be researched and resolved in a sufficiently decisive fashion to prevent an economically plausible SST from being produced and used.

In its concern with scientific unemployment,.FAS shows interests consonant with those of scientific unionism. But FAS cannot be accused of attempted "status preservation" since its lobbying is not directly on behalf of the unemployed, but an effort to bring about conversion of defense-related facilities employing scientists for other purposes. It is doubtful that its approach finds much support among the majority of the currently unemployed. An article in the November 1970 *Newsletter* discusses the alternative macro- and microeconomic approaches to the "serious unemployment problems" which "lie ahead for many of the 600,000 scientists and engineers employed in defense work, or defense-related industries."

> Every time the need to avert widespread technological unemployment hits, observers consider the problem extra-ordinary: the end of a Korean or Vietnamese War or something else unusual. It is evident, however, that extra-ordinary problems are endemic to our social and economic system:

63   FAS *Newsletter,* 23, No. 8 (1970), 3.

continuing methods of dealing with them have to be developed. Even without the extra-ordinary events—as the rate of technology and specialization increases—more and more persons find their education inadequate to continued functioning as a specialist over their lifetime.

The Federation of American Scientists can find a useful role to play in keeping attention focussed—through good times and bad—on the problems of priorities, conversion, and re-education.[64]

The FAS approach is not, therefore, to "politicize" the issue by linking it to a general critique of society, as other critical organizations do, nor does it seek to "individualize" the issue, as many unions might do: its approach might be called the attempt to "intellectualize" the question of scientific unemployment.

(2) The Council for a Livable World

ORIGINS AND RELATIONS WITH THE SCIENTIFIC COMMUNITY. The Council for a Livable World (CLW) was founded in 1962, largely the creation of Hungarian-born nuclear physicist Leo Szilard. Szilard, actively involved in the Manhattan Project (development of the atomic bomb), had later opposed use of the weapon. In an attempt to influence President Roosevelt, he had stimulated Einstein to write his famous letter to the president of March 1945.[65] After the war, Szilard continued active in the attempt to develop international means of control over the new weapons. In 1961 he published his well-known story, "The Voice of the Dolphins," in which scientists find ways of tapping the "superhuman" intelligence of dolphins, and are thereby informed how to prevent international disaster.[66] In 1962 Szilard embarked on a lecture tour of colleges and universities, delivering an address entitled "Are We on the Road to War?" In his speech, he outlined a number of steps governments could take to reduce the antagonistic nature of postures adopted by the United States and the USSR toward each other. The ultimate objective was general disarmament, but the immediate strategy that Szilard proposed embodied financial support for sympathetic candidates for political office. He asked that supporters of the scheme pledge 2 percent of their annual income, in order to build up a "national constituency" for antiwar candidates whose localities could not or would not provide adequate campaign funds.[67] The overall "movement"

---

64 *Ibid.*, p. 2.

65 Smith, *A Peril and a Hope*, p. 28.

66 L. Szilard, *The Voice of the Dolphins and Other Stories* (New York: Simon & Schuster, 1961).

67 Elinor Langer, "Scientists in Politics," *Science,* 145 (1964), 561–563.

would be directed by a committee of scientists and scholars. After the tour requests for money were sent out to potentially interested individuals, largely in universities. Within a few months enough money to commence operations had been received, and a Council for Abolishing War was set up. The council began its work, along lines suggested by Szilard, during the congressional elections of that year.

There are today some 12,000 members—or, more accurately, supporters—of the council, about 40 percent in academic life and 40 percent in professional and business life. The council, responsible to a thirteen-man Board of Directors, is managed by a small group of officers and a full-time director. There are links with FAS, since the president and treasurer (professors Bernard Feld of MIT and Matthew Meselson of Harvard) are also members of the council of the federation. MIT political scientist George W. Rathjens is also on the board of one and the council of the other. Chairman of the CLW board is William von E. Doering, Professor of Chemistry at Harvard and a member of the National Academy of Sciences. Nonscientists are on the board too, such as Roger Fisher (Professor of Law at Harvard), Daniel Aaron (Professor of English at Smith), and James G. Patton (ex-president of the National Farmers Union).

ACTIVITIES OF CLW. The council concentrates its attention on the U.S. Senate, since this body has primary legislative responsibility for the foreign affairs and defense issues in which it is interested. Most of its work is concerned with securing the election of antiwar senators, as envisaged by Szilard, principally by channeling financial aid into suitable campaign funds. The basis of this operation, from its first campaign (in 1962) to that of 1970, is that supporters transmit checks via the council offices to recommended candidates. In the 1962 election $58,000 was contributed, of which $22,000 went to Senator McGovern (who won by only 600 votes). Available funds have built up steadily, until in 1970 over $300,000 was contributed. Support is largely directed toward the smaller states, where it can produce the greatest impact—to Alaska, Nebraska, Utah, Tennessee, and so on.[68] Candidates are not chosen on a party basis, but solely because of their record or attitude on such issues as control of nuclear weaponry. Moreover, a candidate would not be offered support if his opponent had similar views on such issues. Thus, the council maximizes the influence of its funds.

In addition, the council runs conferences and seminars in Washington, "in which it strives to inject new ideas into appropriate political channels, to encourage national discussion of controversial proposals and to facilitate the

68   "Scientists and Senators Against the Arms Race," *Nature,* 228 (1970), 406–407.

involvement in vital national issues of the most knowledgeable and articulate persons outside the government." [69]

The CLW is widely recognized as politically effective, by friends and enemies alike. It has been attacked fiercely from the right, on the grounds of being part of the "Red China lobby," and also on the grounds of pledging recipients of its aid to support its specific policies, which is not the case. In 1970 it was accused by Vice-President Agnew of advocating unilateral disarmament of the United States: [70]

> Perhaps the foremost lobby for unilateral disarmament in the United States is a little known, even less publicized "Council for a Livable World."
>
> Every measure to slash America's military strength seems to have the enraptured backing of this lobby.
>
> This Council holds alarming leverage over some members of the United States Senate.

ISSUES OF CONCERN TO CLW.   Above all, the council aims to halt the arms race and establish international peace-keeping mechanisms. It has fought against ABM and MIRV, supported SALT and U.S. ratification of the Geneva protocol on chemical and biological warfare, sought to end nuclear proliferation and the war in Indo-China. Some idea of its "style" can be obtained from the way in which its views on such issues are expressed.

In its opposition to the Safeguard ABM, the council adopted a "realistic" stance, opposing the ABM system specifically, not the whole concept of nuclear armament or need for a U.S. deterrent:

> It may well be time to recognize that the fixed ICBM has served its purpose, and that in the future we should rely primarily on the mobile sea-based systems for deterrence. This we can safely do.
>
> No government can mount a nuclear attack on the United States with any hope of surviving the retaliatory blow which our Polaris fleet can inflict.[71]

Opposition to Safeguard, whatever its *source,* was *expressed* primarily as objection to the project's technical feasibility, to the ultimate cost ($50 billion or more), and to the inevitable effect upon the SALT talks. "Let us ask whether the ABM program for the defense of Minuteman makes sense

---

69 Council for a Livable World brochure (1970) (this and subsequent quotations © Council for a Livable World, Washington, D.C., 20002, U.S.A.).

70 Speech of Vice-President Agnew quoted in December 1970 mailing of the council.

71 "Memorandum on Secretary Laird's ABM Proposal of February 24th 1970" (CLW, February 1970).

even within the Administration's own frame of reference." [72] Taking the Secretary of Defense's own estimates of the situation, the council's authors concluded that the system would be inadequate if (in the Secretary of Defense's words) "the Soviets deploy a MIRV on the SS-9, improve their ICBM accuracy, and do not stop building ICBMs at this time but continue building them at their present rate"—which they interpret as inadequate under all likely circumstances!

The council objects also to the way in which the "President has come to rely almost exclusively on a small group of officials within his own Administration for advice on national security and defense problems." It appeared to the council, from presidential statements on the "infallibility" of the anti-China defense system, that "the President has not only failed to avail himself of the advice of knowledgeable senators but has also neglected to draw on informed scientific opinion outside the government. No responsible scientist or engineer with experience in military technology would support the concept of an 'infallible' population defense." [73] The same "realism" informs the council's attitude toward CBW (chemical and biological warfare), which it opposed, and on other issues. And, again, there is an objection to the ways in which decisions on such matters are made, and the limitations deriving from them: [74]

> Last April President Nixon ordered a broad review of CBW policies within the Executive Branch. It was a welcome development. However, nearly all the experts on these weapons within the government are military men who cannot be expected to present the President with the fullest range of policy choices.

(3) The Society for Social Responsibility in Science

FOUNDATION AND PURPOSES. In contrast to the organizations discussed above, the Society for Social Responsibilities in Science (SSRS) is an international society, founded in 1949, with its headquarters at Bala-Cynwyd, Pennsylvania. The majority of members are American, but substantial groups live in Western Germany and Japan, and smaller numbers in over forty other countries. The society was founded in the belief that scientists had, and have, a solemn obligation to ensure that application of their work is in the best interests of humanity. Scientists, its founders urged, must stand together to prevent international war and to ensure survival of a seriously threatened humane civilization. Membership was opened to men

72   *Ibid.*

73   *Ibid.*

74   "CBW and the Geneva Protocol: The Choices Just Ahead" (CLW, September 1969).

and women of all nationalities and creeds educated, or working, in physical, biological, and social sciences, who subscribed to the society's purposes. (There is an associate membership for nonscientists.) The society has a twenty-four-man council and composition of this council distinguishes SSRS from FAS and CLW. Although past and present membership has included a number of the great names of international science (Einstein, Pauli, Born, Pauling, Szent-Gyorgi, Yukawa, Luria), the society's council includes few of the "big names" of American science. E. U. Condon is perhaps the only council member famous internationally in the field of "science and politics," and there is only one member of the Harvard/MIT science aristocracy. Perhaps more interesting is the fact that many council members are engineers and applied scientists and half its membership (including President Alice Mary Hilton and Vice President Earl Graham) are not academics. SSRS appears representative of a substantially *wider* segment of the scientific community in America, and this is undoubtedly a consequence of its apolitical stance. The society's orientation is much more compatible with straight "professionalism" than are the orientations ("styles") of other organizations considered here. Its purposes emphasize responsibilities of the *individual,* as do the majority of professional creeds. They are set out as follows:

> The primary purpose of the SSRS is to help scientists and engineers to meet, by constructive means, the central problem of our day: the survival of civilization itself in an age in which the destructive power of military weapons reaches ever more devastating proportions. In this context, the SSRS calls upon every scientist and engineer (1) to foresee, insofar as possible, the results of his professional work, (2) to recognize his personal moral responsibility for the consequences of his work, irrespective of outside pressures, (3) to seek work which seems to him of benefit to mankind and abstain from that which seems to him injurious to it, and (4) to use his scientific and technological knowledge, guided by ethical judgement, to aid government and laymen in the intelligent and humane use of the tools which science and technology provide.[75]

ACTIVITIES. SSRS is not a political organization, in the sense that it does not engage in direct lobbying activities. Its interest in the effects of science upon human life has led it, rather, into essentially educational contemplative activities, as follows:

1. A program of discussion and education within the scientific community, aimed to provide the sort of information upon which essentially moral judgments may be made by scientists.

75 Publicity handout, Society for Social Responsibility in Science (1971).

2. Stimulation of informed public discussion.

3. Promotion of projects of clear humanitarian value, such as development of agricultural and technological tools for developing areas.

4. Inquiry into the logical basis of science as it might affect formulation of an ethical code for science.

5. Employment service for those individuals whose convictions require them to refuse potentially destructive work.

6. Study of the social implications of new technologies such as weather control and automation.

7. Studies of environmental problems and of the ecological consequences of, for example, military activities.

In terms of its activities, therefore, the Society for Responsibility in Science is not really a development *upon* professionalism and, in its emphasis upon *individual* action, it is committed to that remoralized professionalism discussed in the previous chapter. It is this, no doubt, which ensures its acceptability to what must be a less than militant segment of the scientific community.

(4) The Union of Concerned Scientists and "March 4th"

"March 4th" or "march forth" was essentially an MIT activity born late in 1968, initially the idea of a group of graduate students.[76] From the conviction that a threatened or, if necessary, realized total research stoppage might bring a sufficient pressure upon the U.S. government to ensure cessation of the Vietnam war, they evolved the concept of a symbolic one-day stoppage. Gradually, the students involved members of the faculty, first in the physics department and then beyond it. A formal statement was prepared and refined by discussion, with a program of speakers arranged for a day of discussion. For up to three months, a number of graduate students devoted their whole time to preparation of the activities. A joint student-faculty program committee was established, with Professor Hermann Feshbach as chairman, and March 4, 1969, was chosen for the stoppage. As arrangements and discussion developed, however, faculty and student viewpoints diverged. Many students objected to the "timidity" of the faculty and their reluctance to jeopardize their status in the institute. For their part, many faculty members were anxious to avoid total polarization of the teaching staff and potential endangering of the stability and standing of MIT. Differences were crystallized by a preemptive report of the proposed action in *Science.* Subse-

76 M. Eden, "Historical Introduction," in J. Allen (ed.), *March 4 Scientists, Students, and Society* (Cambridge: MIT Press, 1970).

quently, it was agreed that "March 4th" would be sponsored by two distinct organizations—of students (Science Action Co-ordinating Committee, or SACC) and of faculty (Union of Concerned Scientists, or UCS). A statement of intent and objectives, derived from the students' initial manifesto, was circulated among the faculty and widely accepted. It represents the clearest statement, at least from the faculty viewpoint, of what the protest was all about. Ending with an invitation to participate in the stoppage (the word "strike" was avoided by those involved), salient parts of this statement read as follows:

> The response of the scientific community to (the misuse of scientific and technical knowledge) has been hopelessly fragmented. There is a small group that helps to conceive these policies, and a handful of eminent men who have tried but largely failed to stem the tide from within the government. The concerned majority has been on the sidelines and ineffective. We feel that it is no longer possible to remain uninvolved.[77]

They proposed "a critical and continuing examination of government policy" to devise means of converting scientific application to civilian uses, to persuade students of the importance of bringing benefits of science to mankind, to express their objection to ABM and other military projects, and to see whether scientists and engineers could be organized so that "their desire for a more humane and civilized world can be translated into effective political action." A manifesto along these lines was signed, *inter alia,* by heads of the biology, chemistry, and physics departments (Professors Boris Magasanik, John Ross, and Victor F. Weisskopf), as well as by other MIT illuminati such as Chomsky and Luria.

Not surprisingly, there was a good deal of concern over the specific gesture adopted: the research stoppage. Science revolves around pursuit of research activities, and university teachers value such time as they can keep free from their teaching duties and devote to research. Research time is not to be sacrificed lightly. Therefore, the decision to forego a day's research represented an explicit subordination of the scientist's duty to science to his duty to society: this, indeed, is the interpretation put upon the action by Professor Murray Eden in his historical account of origins of the protest.[78] But there was an alternative view among the MIT faculty: that the symbolic gesture should be dropped in favor of a day's discussion. Did this show tacit fear of the political power of symbolism? But the alternative was rejected.

The stoppage was executed, widely reported,[79] and contributed papers

---

77   Faculty statement, *ibid.,* p. xxii.

78   *March 4,* Historical Introduction, p. xviii.

79   For example, in *Science.*

were published.[80] The feeling was that an important new movement had been launched—an alliance across the generation gap of science. But this proved less than the case.

In March 1970 an anniversary meeting was held, without the stoppage, under the sole auspices of the faculty body UCS.[81] The faculty-student alliance had turned sour: the confrontation-minded SACC could not accept UCS's desire for consensus, for cushioning the university itself from attack. Symbolic and instrumental views of politics proved incompatible: agreement over ends was offset by disagreement over means. UCS will not engage in such symbolic gestures as commitment to the "scientific pledge" (not to engage in war research under any circumstances). Of late, the difference in view has been crystallized by a "confrontation" between the more radical SESPA (see below) and UCS.[82] SESPA members feel that UCS has not done enough to cut down MIT's substantial involvement in military R & D. Thus, a SESPA spokesman, Seymour Melman (Professor of Industrial Engineering at Columbia), is quoted by *Science* as faulting liberal faculty members and students at MIT for "avoiding responsibility for the professional character of (their) institution," and for saying that "Pentagon control of research doesn't matter if you are pure in heart and try to do your own thing with the money." [83] The chairman of UCS admits that members might be prepared to carry out weapons research under some circumstances, and so cannot make any pledge.

> There are circumstances when some of us would work on weaponry. We are convinced that now is not such a time. We devote our energies and our talents so that the time may never come.[84]

An affinity with the aims and approach of the FAS has been recognized by UCS, and it has recently affiliated itself with that organization.

(5) Scientists and Engineers for Social and Political Action (SESPA)

FOUNDATION OF SESPA. This organization grew out of failure of the institutions of science to respond to the increasingly apparent political situation of science. More specifically, it developed out of the failure of physicists Charles Schwartz, Martin Perl, and others to persuade the American Physi-

---

80   J. Allen (ed.), *March 4 Scientists, Students and Society* (Cambridge: MIT Press, 1970).

81   Hamilton, "MIT: March 4th Revisited Amid Political Turmoil."

82   T. P. Southwick, "Visitors Ask MIT Faculty to Renounce Military Research," *Science,* 171 (1971), 156.

83   *Ibid.*

84   Letter from Professor Lee Grodzins, *Science,* 172 (1971), 214–215.

cal Society to take a stand on the Vietnam war. The so-called "Schwartz amendment" to the APS constitution was voted down in 1968. SESPA began, as Scientists for Social and Political Action, early in 1969, as Schwartz and others became aware of the extensive concern over issues such as ABM, DoD funding of research, and the war, within the scientific community.[85] The founder members were a group of about one hundred scientists, primarily physicists: "The foundation of SESPA was simply the expression of organizational need of widespread political activity among scientists." [86] As the organization grew, the name was changed to include "Engineers."

It is difficult to find a specific statement of SESPA's objectives and there is little doubt that it serves as an umbrella for much disaffection with the state of science, of America, of the world. But perhaps geneticist James Shapiro's reasons for abandoning science in favor of social and political activity may encapsulate the principal directions of the movement: he is very active in SESPA. Shapiro gave three reasons for leaving research. First, he believes that the results of science (including his work) will be put to "evil use" by governments and large corporations; second, he is not prepared to work under a system which does not permit "the people" to help decide scientific priorities; third, he believes that political solutions are more urgent than scientific ones in alleviating social ills such as pollution.[87]

> But political and social action was not the only dimension. Learning through their struggles SESPA people were beginning to develop analyses of contemporary American Society and of the role of Science and Scientists.[88]

RELATIONS WITH THE SCIENTIFIC COMMUNITY. Membership of SESPA (early in 1971) is probably about three thousand plus scientists and engineers, although circulation of the bimonthly periodical *Science for the People* is around five thousand. SESPA is a federation of local groups in some twenty American cities, principally (though far from entirely) on the east and west coasts. When SESPA was formed, it was agreed that it was to be "a non-organization," emphasizing local activity and geared essentially to maintenance of communication between these local groups. SESPA has no president, no council, no official spokesmen. But those who write in *Science for the People* are younger than those who write on the same issues in *Science,* and a high proportion are women. There is little doubt that SESPA appeals to a very different segment of the American scientific community than those segments attracted to SSRS, FAS, or CLW.

85 "Scientists Promote Political Role," *Science,* 164 (1969), 3875.

86 "SESPA—A History," *Science for the People,* 2 No. 4 (1970), 3.

87 James K. Glassman, "Harvard Genetics Researcher Quits Science for Politics," *Science,* 167 (1970), 963.

88 "SESPA—A History," *Science for the People,* 2, No. 4 (1970), 3.

SESPA ACTIVITIES.   Of all SESPA activities none have attracted more
attention, certainly from the mass media, than its disruptions of AAAS meetings. Thus, reporting the 1969 (Boston) meeting, *Science* carried an article
entitled "AAAS Boston Meeting: Dissenters Find a Forum," [89] describing
some activities at the meeting. Student activists "handed out their own 'Mars
rock' at the NASA exhibit," "held a coronation ceremony for a 4 year old
'Miss Moon 1970' " and distributed "science for the people" badges. Publicity had been sought and won without very much trouble. Discussion meetings were organized, others (official ones) passively interrupted, and resolu-
tions advanced on such subjects as equality for women in science and need
for inquiry into use of herbicides in Vietnam, requiring AAAS to demonstrate its commitment to human welfare by demanding U.S. withdrawal from
Vietnam. The need for the inquiry was accepted, and an AAAS team under
Harvard biologist Meselson did go out, reporting back to the 1970 meeting.
However, a young Harvard astronomer, associated with SESPA, is quoted
as saying that "it did not matter . . . whether the council passed the resolu-
tions, their value was purely educational." Similar activities were mounted
at other major scientific meetings (American Physical Society, American
Chemical Society, Acoustical Society of America, etc.) and, again, at the
1970 AAAS meeting in Chicago.

In 1970 there was an escalation in the scale of confrontation and, even
more than its predecessor, the Chicago meeting was represented by the media
as a manifestation of scientific discord and violence. The British periodical
Nature, perhaps better able to preserve a sense of detachment, recognized
the entertainment value of the proceedings:

> Circus is not too inaccurate a description of a meeting that heard little
> science, sacred or otherwise, but witnessed a succession of theatrical dis-
> ruptions which chairmen of sessions were mostly unable to control.[90]

Glenn T. Seaborg was indicted for the crime of "science against all people"
("You are guilty, Glenn T. Seaborg, of a conscious, major, self-serving and
ruthless role in establishing, organizing, maintaining and developing institu-
tions of science and government for effective rule by the ruling class. WE
INDICT YOU FOR THE CRIME OF SCIENCE AGAINST ALL PEO-
PLE").[91] Edward Teller, "war criminal," was presented with the second
annual Dr. Strangelove Award ("SESPA IS NAUSEATED TO PRESENT
ITS SECOND ANNUAL DR. STRANGELOVE AWARD TO EDWARD
TELLER, in recognition of his ceaseless efforts to follow in the footsteps of

89   *Science,* 167 (1970), 36–38.
90   *Nature,* 229 (1971), 81–82.
91   *Science for the People,* 3, No. 1 (1971), 12.

the great Peter Sellers").[92] (He refused to accept it!) Philip Handler (a "lackey of the ruling class") gave a talk "studded with quotes from establishment party platforms, woeful cries about U.S. science losing world leadership and digs at women, students, and dissenters."[93] Edward Teller reacted by providing himself with five armed bodyguards! Seaborg reacted by refusing to take the platform as SESPA members read out their "indictments" of him.[94] *Nature* seems to have appreciated the truly theatrical nature of the proceedings:

> The indictment, like the other SESPA activities, was intended more as a harmless piece of theatre than a factual criticism, and attacked the offices that Seaborg has held rather than Seaborg the man, being devoid of specific or personal details.[95]

The symbolism seems to have been maintained even in the *responses* to the disruptions: Teller's bodyguard, the "stabbing" of a heckler with a knitting needle by the irate wife of a speaker.

If SESPA's intention was to get across to the eight thousand or so participants, arousing the interest of viewers and newspaper readers, they seem to have been successful. "The theatrical gestures and noisy disruptions may have got the message across, even if nothing else, to a largely hostile or indifferent audience."[96]

SESPA's theatrical repertoire is wider than this. The "Science for the People" button, depicting a clenched fist and a hand holding a chemical flask, is splendidly dramatic, so is the substitution of AAA$ for AAAS. So, too, is the SESPA pledge:

> I pledge that I will not participate in war research on weapons production. I further pledge to counsel my students and urge my colleagues to do the same.

although in a different way, and the picketing of research institutions. But, to be fair, there are other activities of a more practical kind. Boston SESPA has developed a Technical Assistance Program, devoted to provision of tech-

---

92 *Ibid.,* p. 10.

93 "SESPA Tells It Like It Is: Opening Statement AAA$ '70," *Science for the People,* 3, No. 1 (1971), 6–7.

94 It is not necessary to draw any causal relationships, but certainly there was a considerable, almost unprecedented concern at the nomination of Glenn Seaborg (then chairman of the Atomic Energy Commission) as president-elect of the AAAS. "Seaborg Brings Strife to AAAS," *Nature,* 228 (1970), 1023–1024.

95 *Nature,* 229 (1971), 81–82.

96 *Ibid.*

nical information, and sometimes practical help, to community groups. (For example, they helped build an electrical generator when the Edison Company "denied power" to a free community medical clinic.) But, on the whole, SESPA activity is symbolic directed inward to the scientific community. This, after all, is why AAAS must occupy a special place in the SESPA diary. AAAS, far more than any other body, purports to be, and is regarded as, representative of the American scientific community. And yet its leadership is "the scientific elite . . . (whose) prestige and financial security depends upon the maintenance of present institutional forms." [97] According to the SESPA analysis, leadership and membership have totally different interests, a microcosm of society-at-large. At the same time AAAS, unlike most scientific bodies, has a *de jure* commitment to "improve the effectiveness of science in human welfare." But SESPA finds its practical demonstrations of commitment far from impressive:

> Further evidence of the effort being devoted to social action is furnished by the annual financial statement of the AAA$. In 1970, of a total expenditure of 5 million dollars, 15 thousand (0.03%) is reported for Public Understanding of Science and none (0.00%) is reported for Promotion of Human Welfare (or anything resembling that).[98]

SESPA resents what it sees as the divergence between AAAS commitments and AAAS practice. It demonstrates a concern for what is no more or less than normatively prescribed behavior. "Action is necessary to close the gap between the pronouncements and the practices of the scientific community." That an essentially moral crusade should be conducted through a series of essentially symbolic actions is far from unusual. This is basically the nature of SESPA, supported by consideration of issues with which it concerns itself and the style in which pronouncements are made.

ISSUES AND PRONOUNCEMENTS.   An issue in which SESPA has shown a continued interest is the position of women in science. The issue was taken up after a resolution, signed by some hundreds of participants at the 1969 AAAS meeting, had been rejected for publication by the editor of *Science*.[99] Their critique of the current position of women scientists, some of which sounds familiar to anyone connected with Women's Liberation, runs something like this. In the first place, the educational system pressurizes women into a limited number of stereotypical female roles: housewife/mother for preference; next best, for the highly educated, the service professions (teaching, nursing, social work). The few who manage to avoid this "brain-

97   "History of the AAA$," *Science for the People*, 2, No. 4 (1970), 15–22.

98   *Ibid.*

99   "Equality for Women in Science," *Science for the People*, 2, No. 2 (1970), 10–11.

washing" process find themselves opposed by further barriers. Restrictive quotas for women entering graduate school are justified by appeal to the "fact" that women are unlikely to remain long in the profession and "lack the emotional stability and drive to meet the arduous initiation rites of the profession." [100] A later issue of *Science for the People* gives details of one or two specific cases of academic discrimination against women.[101] So those who become scientists generally find it very hard to escape from an imposed position of inferiority in science, "rarely being given their own labs or first authorship on papers, and, the most glaring inequality, being paid less than their male colleagues for equal work." It is not difficult to show that such sex discrimination is contrary to the norms of science and to the principles of the American Association for the Advancement of Science.

> Clearly we cannot "further the work of scientists" while denigrating in so many ways the contributions and potential of women in the profession. Sexual discrimination makes "co-operation among scientists" an ironic platitude.

It would seem that here (as elsewhere) SESPA has hit upon a pathological feature of the social organization of science deriving from that organization's dependence upon the values of society-at-large. This dependence is recognized in their analysis.

> Female scientists do not escape the oppression faced by all women in our society. . . . Such sexual discrimination is no accident. It serves, in a variety of ways, the interests of those who dominate the economy of this country. It provides them with a source of ideologically justified cheap labour, and as a consequence drives all wages down.[102]

At one level of analysis, this is one among many examples of departures from universalism in science (from the norm requiring that access to a scientific career and recognition of achievement in science both be determined by merit, not by other cultural or social factors). At another level, it is a telling example of distortion of the fabric of science by the nondemocratic nature of modern Western society. The same kind of distortion, stemming once more from the uneven distribution of power in society, can be reflected in the substantive nature of scientific advance. Again, the dominance of men in society is an example of this power distribution, and the substantive nature of birth control research is taken as its effect.

100   *Ibid.,* and subsequent quotations.

101   "Discrimination at U. Mass.—Woman Scientist Fights Back," *Science for the People,* 3, No. 2 (1971), 18–21.

102   "Equality for Women in Science," *Science for the People,* pp. 10–12.

How is birth control practiced in our society? It should come as no surprise that in a society where women are the lower caste, birth control is practiced by intervention on the female body. The upper caste, after all, runs the show.

As we go along struggling with the burden of our reproductive system, we might begin thinking as we did when we were very young. "If I could only be free of this oppression, be like a boy. . . ." The brutal answer to this ancient longing is the male scientists' public advocation of the sterilization of women. Should we be surprised? In a death oriented society overkill is the ultimate solution to any problem. And who's going to get sterilized? No males . . . "no one's fooling with my sperm, baby!" [103]

Here, then, is an area of technological development which has not received open discussion, and which exists in a symbiotic relationship with research dictated by social norms. "The power structure in our society is male. The scientists who do birth control research are male." The term "power-penis-potency complex" (PPP) is coined:

The time is long overdue to ask about male contraceptives and male sexual responsibility. Where is it? . . . Djerassi . . . advances three "reasons" for the lack of effort to develop male contraceptives (1) He claims less is known about the reproductive biology of the male than the female. Bullshit. . . . It is definitely a simpler situation (to experiment upon). (2) He says it is easier to experiment on women than on men, because they've got us already through the services of their medical lackeys and the Planned Parenthood Clinics. True. In other words it's the politics of sexism that scientists serve. (3) And of course that old favourite, that classic reason . . . "the male's generally lesser interest in, and greater reservation about, procedures that are aimed at decreasing his fertility." Tst, tst, tst.

Could it be that the whole power complex trembles at any notion of sperm control? [104]

Like the other organizations discussed, SESPA objects to war research and to many restrictions imposed upon scientists in the interests of "national security" by the ruling "military-industrial complex." But SESPA distinguishes itself from these other more "realist" organizations in the phrasing of its objections—as it fully realizes. This comes out clearly in their discussions of the Defense Facilities and Industrial Securities Act, which sought to impose much more rigourous security procedures for workers in a wide range of public institutions (including universities). *Science for the People* reports objections to this bill expressed in a *Science* editorial, and in other ways, by the "liberal establishment," emphasizing the "adverse impact on the

103   "Birth Control in Amerika," *Science for the People*, 2, No. 4 (1970), 28–31.
104   *Ibid.* The reference is to C. Djerassi, "Birth Control After 1984," *Science*, April 9, 1970.

careers of people who might be denied access to research information or research facilities."

> Of course every scientific and technological worker (and everyone else) should oppose this bill, but these "distinguished scientists" should also try and transcend the element of self-interest and protection of special privilege that is evident in their response. After years of holding security clearances and sitting on government councils they feel threatened. . . . Now that their privileges are threatened, will they learn that there is but one struggle, on repression? [105]

Again, this issue is relevant to the discussion of departures from scientific norms, for the SESPA position can be viewed as a critique of formation of an elite in the scientific community.

A final issue, again one preoccupying almost all organizations discussed in this chapter, is unemployment of scientists. Again, SESPA diagnosis and prescription differs from those of other bodies. According to *Science for the People,* treatment of the problem in the scientific press has been in two stages.[106] First, an attempt was made to minimize the problem and explain it away. (This is almost incontrovertible.) Then, following acceptance of the situation, arguments were in terms of an "unfortunate overselection by individuals of certain fields as a result of the Sputnik panic. Now people are seen as having wrong skills. . . . Still the blame is placed with the individual. He picked the wrong job, so he better retrain. . . . If there are no jobs to retrain for, he can take work that requires fewer skills (as *Time,* October 5, 1970, p. 84, well describes), so go the arguments." [107] SESPA's prescription is different. Retraining will be of limited value:

> The only hope for the great majority of unemployed scientists and engineers is not in individual adaptation, but in structural change of the conditions which determine their employment perspectives.

This must imply a political reaction to unemployment: not the acceptance of individual responsibility; certainly not the demand of resumed high levels of defense spending. The liberal argument in favor of "conversion" of defense facilities to civilian uses is not accepted as a very simple remedy. The SESPA argument, a sophisticated one, is that the conversion process will not spontaneously take place as defense funds are cut—perhaps not even "at the margin" (as economists would have it). First, much defense work is

105   *Science for the People,* 2, No. 3 (1970), 4–5.

106   "Unemployment of Scientists and Engineers," *Science for the People,* 2, No. 4 (1970), 5–9.

107   *Ibid.*

done under contract in private industry. Nondefense alternatives will appeal to industrialists only if potential profits are comparable with those from defense work. This implies federal subsidization, as at present. This would require a high and inflationary scale of expenditure. Second, urgent public needs are of the kind more appropriately met by public authorities than defense enterprises. But, to do this, government would have to expand its manpower (and other resource requirements) to a scale which would bring it into conflict with industry. This is unacceptable under U.S. capitalism. Finally, the whole disciplinary and specialist structure of the scientific community is closely geared to military needs and the size and growth rate of the military R & D budget.

THE MOVEMENT IN CONTEXT

In this chapter, I have attempted to categorize certain responses of the scientific community to a situation in the face of which traditional attitudes and institutions seem inadequate. For a complex of reasons, scientists in many countries—the United States and Britain were taken as examples, but there are others—are disaffected and demoralized to a substantial degree. I suggested that the relevant factors have generally been held to stimulate unionization among professionals (deteriorating status, declining salaries and employment prospects, increasingly inadequate working conditions, lack of participation in relevant decision making, etc.). Subsequently, I suggested that the consequent "unionization" movements can be categorized as of two kinds, depending upon their preferred emphasis upon instrumental or ideological grievances. Case studies were intended to illustrate the instrumental nature of the reaction among British scientists and its ideological nature among Americans. In this final section I should like, first, to adduce further evidence for regarding unionization in Britain and politicization in the United States as variants of the same phenomenon and, in conclusion, to emphasize once more the relevance of all this for the sociology of science.

In the first place, however, let me make clear that, although I have described what I regard as the major responses of the British and American scientific communities to the current situation of science, this does not imply the absence of ideological concern in Britain or of instrumental concern in the United States. Such is not the case. In Britain there is a lively British Society for Social Responsibility in Science (BSSRS), an organization with more radical aspirations than its American cousin. BSSRS began toward the end of 1968, launched with an impressive list of scientific sponsors (a number have since been "frightened off" by the leftward movement of the society). Membership has increased from under six hundred in the autumn of

1969 to over twelve hundred in the autumn of 1971 (about 83 percent are scientists or science students; 75 percent—including about 13 percent under-graduates—are in higher education).[108] President of the society is Nobel laureate professor Maurice Wilkins, but the committee consists largely of younger and less established natural and social scientists. Although the society has conducted a number of valuable inquiries and held various discussion and seminar meetings of some interest,[109] it is probably true to say, with *Nature,* that it has not yet found itself an appropriate niche within either society-at-large or the scientific community.[110] Thus, while BSSRS has given visibility to the cause of radical science in Britain it is, in my view, significant that so much discussion of relevant issues turns rapidly to the United States and Vietnam. The easy communication between the two countries, the high degree of awareness of and interest in the more substantial radical science movement in the United States may be inhibiting development of an indige-nous British movement. This example of ideological "unionization" in Britain finds a parallel instrumental movement in the United States. But this seems no more than incipient: employment worries and status considerations bal-ance one another to an extent that crystallization is inhibited.[111]

Why the situation is as it is—why more unionization in the United Kingdom and politicization in the United States—is difficult to say. Perhaps the answer is to be sought in the relative disaffection of industrial and aca-demic scientists, for unionization is principally a reaction of the former, politicization of the latter. Perhaps academic liberals and radicals in the United States have more reason for disaffection than their United Kingdom counterparts—or perhaps they find "insider" politics less successful. It is less plausible to assume that the converse is true of industrial scientists, and some notion of the relative significance of "professional standing" and salary for each group of industrial scientists would have to be introduced. In any event, our discussion of the relationships of these interrelated movements to the social structure of science does not depend upon resolution of this issue.

Let us return now to the main point: the relationship between the unionization movement in the United Kingdom and the politicization move-ment in America. If it were possible to accept, with Prandy for example,[112]

108 J. Gretton, "Is Science Neutral?" *New Society,* September 9, 1971, p. 448.

109 W. Fuller (ed.), *The Social Impact of Modern Biology* (London: Routledge & Kegan Paul, 1971).

110 "Social Irresponsibility in Science," *Nature,* 229 (1971), 513; see also J. Rosen-head, "The BSSRS: Three Years On," *New Scientist,* April 20, 1972, p. 134.

111 Deborah Shapley, "Unionization: Scientists, Engineers Mull Over One Alterna-tive," *Science,* 176 (1972), 618–621.

112 Prandy, *Professional Employees.*

that unionization (specifically, of scientists and technologists) was necessarily a consequence of an emergent class consciousness, it would seem to
follow that any such unionization has an inevitable political dimension. For,
after all, class consciousness has been an important basis of political division.
But this is not the case. As Lockwood argues: "Concerted action is a function of the recognition by the members of the occupational group that they
have interests in common; class consciousness entails the further realization
that certain of those interests are also shared by other groups of employees." [113] And this is borne out by our cursory examination of the union
movement among British scientists and engineers. So there is no *a priori* relationship between unionization and politicization. Instead, I shall adduce
three pieces of evidence which are strongly indicative.

In the first place, consider the *issues* with which the two groups are
concerned, irrespective of language used or strategies advocated. Fundamental to the "unionists" (for want of a better term) is an involvement with
employment prospects for scientists and engineers. Although unions will ne-
gotiate with individual employers, the government's central role in determining the job market (in both United States and United Kingdom) is so
well known that a political stance is almost inevitable. A second major issue
in collective bargaining by professionals is negotiation of conditions of work.
The academic employees of the City University of New York, for example,
in 1969 entered into agreement with the authorities covering facilities, appointment and reappointment procedures, teaching loads, and so on. A major
stimulus to unionization of professionals is the attempt to secure a measure
of control over work conditions (for scientists provision of research facilities
and time for research figure prominently). A similar concern with the process
of decision making, albeit transposed to a national and nonpersonal level, is
central to the dissension of such bodies as FAS. Indeed, this issue, together
with the misapplication of the results of science, is the major concern of
such organizations (although SESPA is a little different). FAS, UCS, and the
others speak of the immoral uses to which science is put, especially in the
service of the military. Yet this concern, as distinct from strategies employed
in giving it expression, derives from the claims to professionalism of science.
This is clearest of all in the case of SSRS, which also gears its actions to
those compatible with a purely professional orientation. Unionization, too, is
an obvious outgrowth of the professionalism of science (although in its practical rather than moral guise), while the transformation of professional bodies
is an important mechanism of unionization.

The interrelatedness of the substantive sources of the two movements

113 David Lockwood, *The Blackcoated Worker* (London: George Allen & Unwin,
1958), p. 137.

is my first piece of evidence. The second, a somewhat skeletal argument, is as follows. It could be asserted, with a degree of plausibility, that, while radicalization is a movement of the left, unionization of British scientists is likely to be largely a status-conscious movement of the right. How can these contrary tendencies be reconciled? This was a problem facing those seeking to explain the politicization of the European intelligentsia in the 1930s. Kotschnig wrote of demoralization of the unemployed graduates in Europe. He saw in the 40,000 to 50,000 unemployed German university graduates the spearhead of Nazism, and he quotes Reinhold Schairer:

> The less these graduates are able to find a natural expression in their chosen professions, the more will their energies concentrate on the realization of political aims, which promises by way of radical changes to assure them of a right to live and a right to work.[114]

Wood, on the other hand, sought to explain the movement to communism of British intellectuals at that time, and in his study is found the unifying feature. In general, he suggests, the reaction of an intelligentsia to depression is a commitment to *action:* "Action became a value in itself." [115] A story recounted of the economist John Maynard Keynes is illustrative. Finding one of his proposals rejected by the government, Keynes began to advocate the opposite policy with equal vehemence, remarking, "The impossible thing is to do nothing. If you won't do what I think is right, at least you should do the opposite." [116] Why communism in Britain then, rather than fascism as elsewhere? This becomes almost a subsidiary issue; Wood answers in terms of the liberal backgrounds of British intellectuals. Communism, unlike fascism, could be justified in moral terms acceptable to the liberal conscience. It is, in fact, interesting that many prominently radical left (or Communist) scientists seem to arrive at their political positions via a *scientism* more akin to fascism (than to the communism to which it led them) in its pure form.[117]

My final evidence is that, contrasted to the rigidity to which many British scientific institutions seem prone, transformations between political and union orientations seem relatively easy. Specifically, consider the history of the (British) Association of Scientific Workers (AScW). AScW, founded in 1918 as the National Union of Scientific Workers (NUSW), has swung between *three* rather different conceptions of its task. Its initial emphasis, as its name implies, was upon its role as a trade union:

---

114 R. Schairer, "Akademische Berufsnot," quoted by W. M. Kotschnig, *Unemployment in the Learned Professions* (London: Oxford University Press, 1937), p. 175.

115 N. Wood, *Communism and the British Intellectuals* (London: Gollancz, 1959).

116 *Ibid.*, p. 107.

117 *Ibid.*, pp. 127–132.

> The objects of the Union are twofold: they are concerned with the part to
> be played by science in the national life and with the conditions of employ-
> ment of scientific workers.[118]

Here we have two kinds of union orientation: ideological and instrumental.
In 1918 NUSW had 500 qualified members, and by 1922 about 800 (or
16 percent of an estimated potential membership of 5,000). In an appeal
for members published in *Nature* in the early twenties, signed by 21 eminent
scientists (including J. B. S. Haldane, Sir William Bragg, the editor of *Nature*,
Sir Richard Gregory, and H. G. Wells), a more representational conception
of the union's role was propounded.[119] In fact, in 1927 NUSW changed its
name to the Association of Scientific Workers and deregistered as a trade
union, hoping to reconcile professional and political interests. Membership
did not rise through the early thirties but, by the end of the decade, the situation and the role of AScW had changed. It had, in fact, been "taken over"
by a group of more radical scientists, in Cambridge and elsewhere, including
J. D. Bernal, P. M. S. Blackett, Lancelot Hogben, Joseph Needham, and
others.[120] The professional orientation was discarded, and the association
became a pressure group for the "planned" approach to science which its
leaders observed in the Soviet Union. J. D. Bernal, (chairman of the executive) writing in 1939, described its function in his classic manifesto of the
movement, *The Social Function of Science*.[121]

> The basis of the Association is now recognized to be twofold: one, pro-
> fessional and individual concern with preserving and improving the condi-
> tions of employment of its members, and establishing the status of the
> "scientific worker" as in some way similar to that of the doctor or the
> lawyer; the other concern is with the whole position of science in society.
> The two are closely linked.

(Which is, of course, my argument here.)
For a time AScW was able to preserve this complex conception of its
role, but the arrangement did not survive the war. In 1940 eligibility for
membership was widened, and a new section for those qualified at the technician level was introduced. At the same time it was reregistered as a trade
union. In 1942 it affiliated with the Trades Union Congress, and membership continued to rise through the 1940s: from 2,000 in 1941 to 15,000 in

118   Bernal, *Social Function of Science*, p. 400.

119   W. H. G. Armytage, *Sir Richard Gregory, His Life and Work* (London: Macmillan, 1957), p. 109.

120   P. G. Wersky, "British Scientists and 'Outsider' Politics, 1931–1945," *Science
Studies*, I (1971), 67–83.

121   Bernal, *Social Function of Science*, pp. 401–402.

1948. But, increasingly, the association turned toward the problems of the larger part of its membership—the technicians—and at the same time away from ideological concern with questions of national policy. In a pamphlet published in 1944, AScW stated its then current philosophy: [122]

> It is concerned, as one of its primary objects, with maintaining and improving the position of its members in their employment as scientific and technical workers.
>
> . . . we approach all problems from a trade union point of view. We are not drawn off into side issues, as for example into an abstract wrangle as to whether planning would hamper science and scientists.

Throughout the fifties the salary question was paramount: in 1962 it engaged in its first strike, over the salaries of a group of laboratory assistants.[123] Recognizing its essentially technician orientation at the time, AScW merged with the technicians' union ASSET in 1968 to form what is now ASTMS.

Finally, I should like to recapitulate my reasons for seeking to relate the subject matter of this chapter to the sociology of science in its more usual, narrower conception. Two reasons are at least implicit in what has gone before. In the first place, stratification of the scientific community, stimulating departures from the norm of universalism, is very dependent upon the powers over allocation of resources delegated to the elite by modern government.[124] Thus, the rank and file's attempt to secure for itself a measure of influence may be regarded as a check on this imposed "politicization." Therefore, it represents an attempt to reimpose normative behavior. Second, the issues concerning the pressure groups are indicative of just such a relationship. By now it should be clear that applicability of scientific results is important to the resources made available to it: thus, concern for applicability (characteristic of the organizations discussed here) is an important adjunct of the advancement of science. There is another reason for trying to establish this relationship, and one fundamental to treatment of this and the preceding chapter. Scientific unions and pressure groups have developed because of the inability of older institutions of scientists to respond adequately to changing circumstances. To at least a first approximation, the process has much in common with the growth of professional associations, a process which for many institutions of science related to fundamental differentiation of scientific disciplines.

And, finally, there is perhaps the most powerful argument for regarding these and other grass roots manifestations of disquiet within the scientific

---

122 *The A.Sc.W. and Other Bodies* (pamphlet) (London: AScW, 1944), p. 1.
123 Prandy, *Professional Employees*, pp. 139–140.
124 This is discussed in detail in the next chapter.

community as an essential part of the sociology of science. *Systematic* sociological concern for the relationships of science and society has been limited
largely to discussion of the birth of modern science: to the emergence of
science as a semiautonomous social institution in Western society. Put another way, what were the consequences for an embryonic science of the
emergence of powerful groups in society whose values were similar to those
of science and to whose interests the growth of science was complementary?

The situation developing today is perhaps the very opposite of this,
and it is not hard to see the effects of growing scepticism and antirationalism.
It would be an exaggeration to say that these symptoms toll the death knell
of modern science. Yet in a sense their effects upon the scientific community
of today, some of which we have examined, are as relevant to a full understanding of science as a social institution as are the effects of those very different social circumstances which developed in Western Europe three hundred years ago.

chapter 6
Scientists and Government

THE ISSUES

In his essay on the role of the intellectual in public bureaucracy, published in 1945 and reprinted in his *Social Theory and Social Structure,* Robert K. Merton attempted to distinguish the problems faced by those intellectuals (primarily social scientists in his analysis) seeking to influence public policy from within the government machine from the problems faced by those operating from without.[1] The dilemma of the intellectual "who is actively concerned with furthering social innovations" may be expressed as "he who innovates is not heard; he who is heard does not innovate."[2] The reform-minded intellectual who prefers to work from within may find that the bureaucratic forces pushing him into conformism, pushing him toward no more than comparison and evaluation of a limited number of preselected options, are stronger than he had realized. The intellectual slips into the role of technician, a position made easier for him by its apparent avoidance of any necessity for value judgments on his part: that is, it is "supported by the occupational mores of the intellectual."[3]

1 Robert K. Merton, "The Role of the Intellectual in the Public Bureaucracy," chap. 7 of *Social Theory and Social Structure* (New York: Free Press, 1957).

2 *Ibid.,* p. 217.

3 *Ibid.,* p. 219.

A few later writers have sought to expand this "insider-outsider" categorization of intellectuals' relations to the power system. Thus, Coser examined four possible forms of relationship through historical precedent, while admitting the existence of further variants.[4] Intellectuals, for example, may seek to attain power, as did the French Jacobins and the early Bolsheviks. Both cases suggest that, although intellectuals may attain and retain power "during brief periods of revolutionary exhilaration and upsurge," they are rarely able to hold it when the need is for continuing application of routine policies.[5] Second, intellectuals may seek to advise men of power, guiding their policies, as did the British Fabians (the Webbs were prime movers) or Roosevelt's Brain Trusters. Third, they may serve men in power, holding up standards of absolute morality with which to compare politicians' words and deeds: the Dreyfusards in France and Abolitionists in the United States are cited. Finally, Professor Coser suggests, as a distinct role, turning abroad: idealization of some foreign sociopolitical system (such as the Soviet in the 1930s, the Chinese today).

Many of these roles depend for their existence upon the specific intellectual concerns of those occupying them. Intellectuals with a professional or ideological interest in social or economic change may seek direct or indirect power in order to put their ideas into practice and affect a wide range of policies. Such was the case with the Bolsheviks and the Fabians. Significantly, the relationships of natural scientists to the power system are more limited. Very few scientists, in the modern sense of the word in English, have sought direct political power. Scientistic schemes with scientists in control of society have embodied intellectual constructs not of scientists but of social philosophers such as Auguste Comte. Although scientists have attempted to influence political decision making almost since professional science emerged (in Britain, on a substantial scale since the mid-nineteenth century), this has been largely to ensure survival and advancement of science, not to secure implementation of any broader system of social change. The large-scale involvement of scientists in postwar bureaucracy has not solely been a consequence of the scientists' promptings, but involves realization of the mutual need of science and state for one another. With the partial exception of Marxism, no political systems have looked to the sciences for legitimation, and the scientists *qua* scientists have not functioned as critics of the power system.

In brief, then, natural scientists do not have the same multifarious relationships to government as do intellectuals in general. Indeed, it could be argued that prior to Hiroshima there was only one form of relationship with any political significance. The exigencies of war had seen scientists drafted in substantial numbers into the service of their countries, both as researchers

---

4 Lewis A. Coser, *Men of Ideas,* Pt. II (New York: Free Press, 1970), pp. 135–233.

5 *Ibid.,* p. 138.

and advisors. The growth of this technical-advisory role within government was not matched by parallel development of other relationships to the political system—before Hiroshima. Since that time, when grass roots opposition to the exercise of power developed within the American scientific community, large numbers of scientists *qua* scientists have taken upon themselves an essentially critical role. This is especially the case in the United States, and manifestation of this critical function was discussed in the previous chapter. It is not too great an oversimplification to distinguish but two forms of relationship of scientists as scientists to government: scientists as experts, and scientists as critics, corresponding to Merton's insiders and outsiders. But it is too gross an oversimplification to ignore the "fine structure" of these roles. I have sought to distinguish between instrumental and symbolic approaches to the acting out of the critical role in the United States, and similar complexity characterizes the expert role. In order to understand it, one must examine the constraints under which scientists function in the public bureaucracy, and their responses to these constraints.

But I think we must do more in order to fully understand the scientist's role as expert in government. In attempting to grasp the political functions (and, more fundamentally, the political attitudes) of intellectuals, sociologists have explored individuals' relationships to the social system. They have contrasted the middle-class backgrounds of lawyers, often functioning as defenders of the existing order, with the marginality of social scientists and erstwhile revolutionaries. In seeking to understand the more specific political roles of scientists, we may adopt a similar perspective. After all, they hold political views, and it may be that the most vehement critics of Republican, Conservative, or Gaullist policies are significantly different in background from those who are in broad agreement. But there is an alternative perspective, which in my opinion offers a more profitable vantage point for observation of the political roles of scientists *as* scientists. Unlike intellectuals of left and right, who may be united by passage through a similar educational system (but perhaps by no more than nationality), scientists—whether acting the roles of expert, critic, or observer—are united by much more. They are linked, after all, by the *consensual* nature of their knowledge and the *communal* nature of their profession, to an extent not equaled by those engaged in other intellectual pursuits. The perspective of the sociologist of science suggests another viewpoint: the political roles of scientists might be examined in terms of the statuses of their actors within the scientific community. That the role of the scientist as expert is a valid concern of the sociologist of science is given added weight by the fact that, as I hope to show, the (political) value of his advice is frequently dependent upon his relationship to the scientific community.

The main focus of this chapter is upon the scientist within bureaucracy. How are scientific advisors chosen? What departures from prescribed normative behavior are involved in their work, and what responses do conflicting demands produce? What are the implications of this expert role, and the ways in which incumbents are chosen, for the scientific community? In succeeding sections, therefore, I shall examine the tasks set scientist-advisors by governments, the choice of advisors, and the problems of advice giving. Finally, I will turn to the broader sociological perspective of the scientific community.

THE NEEDS OF GOVERNMENT

There are various ways to classify the work of scientific experts in government. Wilensky, discussing the roles of experts-in-general in bureaucracies, distinguishes three on a functional basis: "contact men," "internal communications specialists," and "facts and figures men." [6] These are standard staff positions found in many organizations, and responsible for, roughly, establishment and maintenance of good relations with relevant publics, markets, and constituencies; maintenance of good relations between different groups within the organization; and provision of data for decisions. It is principally into the latter category that scientists fall, and I shall devote space later to consideration of the kinds of information which they provide. But there are other ways than the functional of categorizing the work of experts in government: one might, for example, distinguish various areas of policy making. Some years ago dean Harvey Brooks distinguished between "science for policy" and "policy for science," and such a distinction has proved a useful one, at least until the end of the 1960s. The former refers to formulation of policy within the substantial range of government activities to which research can contribute: defense, health, transportation, and so on. The latter refers to formulation of policy specifically for promotion of science, to some extent "for its own good." A classification along these lines is possible. Schooler, for example, has categorized involvement of scientists in policy making by area, distinguishing degree of influence of experts over decisions. His classification is given in Table 6–1.[7] A final alternative, frequently employed in the international comparison of science policy processes, is essentially structural.[8] Accounts of this kind, showing the relationships of various formally constituted advisory and executive bodies (and rarely seeking specifically to evaluate degrees of influence) may gain added depth by use of historical analysis. How did a given structure develop, and what rationale was offered for structural change?

6 H. L. Wilensky, *Organizational Intelligence* (New York: Basic Books, 1967), pp. 10–19.

7 Dean Schooler, Jr., *Science, Scientists, and Public Policy* (New York: Free Press, 1971).

8 See, for example, the national science policy reviews of OECD and UNESCO.

The overall picture emerging is of an apparatus of growing complexity in all countries, with scientists increasingly involved in policy making as the relevance of science to governmental objectives becomes more and more apparent. It becomes clear too that the exigencies of war have played a crucial role, not only in stimulating scientific effort, but also in triggering the mechanisms for exercising surveillance over this effort. The 1914–1918 war, cutting off the United Kingdom both from the universities and the industrial products of Germany, was substantially responsible for creation of the Advisory Council for Scientific and Industrial Research and its executive department, overcoming previous rivalries of established departments.[9] Similarly, the United States owes its high degree of interpenetration of government and science to requirements of World War II. Thus, an attempt to create a scientific advisory body in the early 1930s foundered, in spite of the New Deal's general sponsoring of outside experts:

> The Science Advisory Board . . . was a failure of the bearers of a tradition of pure and disinterested science, grown up in an environment in which government had played a relatively small role, to adapt themselves to a new situation in which government sought their assistance and offered them its support.[10]

It foundered upon the scientists' inability to recognize and come to grips with a perennial difficulty of science advisors: members of the Science Advisory Board "wanted politicians to appreciate the needs of science but . . . had little interest in the desires or needs of the politicians." The scientists had yet to develop the important capacity for dual loyalty—to government and to science. A further source of conflict at that time was the resentment of the National Academy of Sciences, chagrined at what it regarded (quite rightly) as an encroachment upon its monopoly of scientific advice.

The necessities of war first awakened scientists, then governments, to the need—no longer of science for government—but of government for science. The birth of their famous *affaire* may almost be dated at August 2, 1939, when Einstein, the most prestigious scientist in the world, wrote his famous letter to Roosevelt urging him to finance an atomic weapons research program.[11] Of later developments—the Manhattan Project, OSRD, NDRC—

---

9 I. Varcoe, "Scientists, Government, and Organized Research in Great Britain 1914–16," *Minerva*, 8, No. 2 (1970), 192–216; R. M. MacLeod and E. K. Andrews, "The Origins of the DSIR," *Public Administration*, 48 (1970), 23–48.

10 Lewis E. Auerbach, "Scientists in the New Deal," *Minerva*, 3, No. 4 (1965), 457–483.

11 D. S. Greenberg, *The Politics of American Science* (London: Penguin Books, 1968), chap. 4, "The War Born Relationship."

TABLE 6–1

Summary of Policy Types, Processes and Arenas

Low-Influence Policy Processes

Self-Regulative Policy Arena (private sector or group determines form and content of government programs or actions affecting that sector).
Policy Type = Extraction.
Social Redistributive Policy Arena (redistribution of rights, skill, or wealth to deprived groups in order to increase their social respect and relevance; government, "haves," and "have-nots" as major actors).
Policy Type = Social.
Governmental Redistributive Policy Arena (government's own action to internally reorganize and redistribute roles, functions, and power).
Policy Type = Organization.
Extranational Policy Arena (nondomestic groups as major benefactors of government actions seen as "distributive" and "redistributive").
Policy Types = Disarmament and Arms Control, Foreign Aid, and Foreign Political (Diplomacy).
Distributive Policy Arena (government distributing benefits to groups not in competition; logrolling; no choice or zero-sum situations; one group's claims do not affect another's claims).
Policy Types = Agriculture, Transportation.

Moderate-Influence Policy Processes

Regulative Policy Arena (government as actor establishing rules for groups in competition with one another and with other sectors in society).
Policy Types = Antitrust, Pollution, Transportation Safety, Trade and Balance of Payments, and Conservation.

High-Influence Policy Processes

Economic Management Policy Arena (government as actor behaving in a regulative manner for the general interest; government shifts benefits among groups to promote larger interest of whole economy; originally seen as redistributive, now regulative).
Policy Type = Fiscal and Monetary.
Communal Security Policy Arena (government distributing collective and common benefits to entire society; benefits both tangible and intangible).
Policy Types = Health, Weather, Defense and Deterrence, and Weapons.
Entrepreneurial Policy Arena (government produces benefits it produces itself and distributes widely; government with its own vested interest; government acting as "entrepreneur" in the business of "manufacturing" or "producing" a product).
Policy Types = Space, Science.

SOURCE: Dean Schooler, Jr., *Science, Scientists, and Public Policy* (New York: Free Press, 1971) p. 36. Copyright 1971 The Free Press.

much has been written, and I need not repeat it here.[12] Suffice it to note, though, that postwar demobilization brought with it "deflation": inflated scientific structures and complements became less desirable. But planning appropriate forms of federal involvement in postwar science began almost at once, thanks to Vannevar Bush and others, and the blueprint for what was to come was largely laid out in Bush's famous report *Science, the Endless Frontier.*[13] The burgeoning advisory apparatus dates from the Steelman Report of 1947, recommending creation of an interdepartmental committee on scientific research and development, composed of relevant bureau chiefs, with a secretary attached to the White House staff.[14] President Truman acted on this suggestion immediately, but the committee proved to wield little influence. More recent history shows the upgrading and expansion of this advisory function, and in 1951 a Science Advisory Committee, attached to the Office of Defense Mobilization (ODM) in the executive office of the president, was created. The access of this body to the president was via the director of ODM. But after Sputnik, in 1957, the post of special assistant was created, and the Science Advisory Committee was given direct access to the president. In 1959 a Federal Council for Science and Technology was established, composed of the special assistant and policy-rank representatives of the research-performing government agencies. The immediate postwar "demobilization of science" was gradually reversed as new demands were made of and by scientists.

But let us return to the *function* of scientists within the bureaucracy: What do they do? It is possible to give a relatively uncomplicated answer to this question only by restricting ourselves to one area of policy making, and the process of allocating resources to science is perhaps most relevant.

All governments of industrialized countries pay for a substantial volume of basic or "curiosity-oriented" research with no immediate external function or application envisaged. Much of this research is carried out in universities, though frequently with government funds, and a number of kinds of specific institutions have developed in which government research may be carried out under academic auspices. Traditionally, allocations for such work have been the responsibility of a wide range of government agencies (National Science Foundation, National Institutes of Health, etc.) and others have taken it upon themselves to support basic research for other reasons (NASA, DoD, AEC, etc.). In Britain it has been the traditional responsibility of the semi-

12 Robert Jungk, *Brighter than a Thousand Suns* (London: Penguin Books, 1964); Leslie Groves, *Now it Can be Told* (New York: Harper, 1962); James P. Baxter III, *Scientists Against Time* (Boston: Little, Brown, 1946).

13 Greenberg, *Politics of American Science,* chap. 6.

14 J. Stefan Dupre and Sanford A. Lakoff, *Science and the Nation* (Englewood Cliffs, N.J.: Prentice-Hall, 1962), chap. 4.

autonomous research councils, five bodies responsible since 1965 to the Secretary of State for Education and Science. In other countries centralization is even more complete and, instead of coexistence of five councils considering largely complementary areas of science, France and Canada, for example, boast but one "research council" covering all fields of science (the Centre National de la Recherche Scientifique and the National Research Council, respectively).

Each body, irrespective of country, includes a hierarchy of scientific committees. At the lowest level, these committees will be concerned mainly with evaluation of specific proposals for research and applications for grants from university teachers and others. Committees are usually distinguished on a disciplinary basis: chemistry, biology, sociology, and so on, and grant applications referred to the appropriate committee. Proposals not falling within the mandate of any one committee are at a disadvantage: inter- and multidisciplinary project proposals may be shuffled from one committee to another in the search for the most appropriate home. Since such work has little appeal for main-line practitioners in a discipline (whether because they regard it as irrelevant to their interests or deviant from the discipline), they are unlikely to wish to spend money available for their field upon it. Thus, formation of a grant-giving committee specifically for a new field is a basic aspect of its institutionalization.[15] The work of members of the grant-considering committees consists largely of evaluating one proposal against others competing for the same funds. All proposals will fall within the discipline of the committee members, and they are expected to judge proposals solely in terms of intellectual merit and the individual's likely capability of carrying them out. In general, scientists will be autonomous in the exercise of this function, little different from other such "performance evaluation" tasks.

But at the top of the hierarchy—and the notion of hierarchy is important—both nature of the task and degree of autonomy in formulation of judgment are very different. At these higher levels, scientific committees may be called upon to assist in determination of resources to be made available to specific fields of science, or to individual research-supporting or -performing agencies involved with a number of fields of science. What, for example, are the continuing commitments of the agencies? What are the various rates of cost inflation? How many new scientists must be trained in research? What interesting new projects do the agencies have on the drawing board? These criteria, plus an understanding of the general financial situation, must inform the thinking of high-level committees on such issues. Nec-

---

15   W. O. Hagstrom, *The Scientific Community* (New York: Basic Books, 1965), pp. 204–206.

essarily, overall allocation processes of this kind devolve upon the political decision makers: the scientists' role is essentially that of advocacy.[16]

As one ascends this hierarchy of decision making, the relevance of purely internal (or intellectual) criteria for the decision-making process decreases, and that of extrinsic criteria (economic, financial, social, political) increases. As suggested, whereas at the lower levels the judgments are, quite properly, based upon the intrinsic worth of one project (compared to another not dissimilar project), higher level committees must use their judgment of the potential social, political, economic (or other external) implications of whole fields of science: at the same time, the autonomy they are permitted in acting upon their judgment is reduced. Here is a central problem of the science advocate role: the politician does not regard it as appropriate for the scientist to exercise autonomous judgment on questions of social or economic principle. Though the scientist may be qualified to draw up the balance sheet, setting out potential gains and losses deriving from the preferential support of one or another field of science, the politician must determine the relative weight to give categories in terms of which the balance sheet is constructed. The growing belief that policy *for* science must increasingly reflect social priorities, apparently growing even among scientists concerned with policy,[17] results in subordination of this area of policy making to these external pressures.

So we must recognize, with Schooler, that levels of scientific influence vary between areas of policy, as well as vertically within one area (e.g., allocation of resources to basic science). Schooler attempted to delineate factors serving to determine the relative influence of scientists in different areas of policy and his very comprehensive list is of some interest here.[18] He distinguishes three classes of factors: "exogenous," "endogenous," and "participational." *Exogenous factors* include visibility (openness of policy process); degree of support for the idea that the scientific viewpoint is relevant; urgency in making a decision (i.e., whether there is time for consultation); degree of development of policy area; extent of consensus among the political leadership; involvement of outside interest groups other than the scientific (e.g., business); and level of interest in policy area of the political

16 The U.S. system of budgetary control, with its division of powers, allows for presentation of such arguments in a number of forums. Apart from any influence which the Office of Science and Technology, the President's Scientific Advisory Committee, or the Federal Council for Science and Technology may have had with the budgetary office in the executive branch, congressional hearings allow for further advocacy by scientists.

17 For this "new thinking" see *Science, Growth, and Society* (a report by a group chaired by dean Harvey Brooks of Harvard) (Paris: OECD, 1971).

18 Schooler, *Public Policy,* pp. 40–57.

leadership. *Endogenous factors* include nature of the relevant scientific field
(physical scientists and engineers are more acceptable than social scientists);
degree of specialization of the field; scientific content of the policy area (in
particular, its relationship to technological change); comprehensibility of the
science involved; presence or absence of nonscientific expertise; presence or
absence of disagreement among scientists over proposed policies; apparent
competence of scientists in policy; degree of interest of scientists. *Participational factors* include scientists' formal positions in the decision-making structure and their access, whether formally or informally, to political decision
makers.

It follows from all this that the policy toward a specific area of research
will generally affect the influence of its practitioners, since the former will in
some measure determine the degree of visibility of the field and its advancement, and will be taken as indicative of its status by officials. Such factors
are relevant to sociological analysis of the scientific expert's role. How does
he view his role? How does his occupancy of the role depend upon the degree
of influence which he brings to bear? How does he seek to maximize his influence? Indeed, there has been no such analysis, whether from a reluctance
to challenge the hegemony of political scientists in analyzing such issues, or
whether their sociological relevance has not been perceived, or whether because of empirical problems of data collection, I do not know. In my view the
analysis of the scientist's role within government is of prime importance for
a comprehensive sociology of science. In the first place, scientists accept
service upon these committees as a legitimate, desirable *reward* for scientific
achievement. To be barred from such participation, for political or other
reasons, is regarded both as an infringement of the autonomy of science and
a setback to the individual's career. In the second place, such activities are
no longer peripheral, but central to the performance of basic research in some
areas. Where, for example, very large equipment is involved, resource allocation procedures must be vital to effective prosecution of research. The
task of carrying out this and other refereeing tasks is central to the social
system of science. But, generally, the scientific advisor works under external
constraints not imposed upon status judges of other kinds, as referees of
papers. Recognizing the importance of these evaluative roles to the social
system, Merton and Zuckerman recently carried out a detailed study of the
referee system: [19] it is my belief that an analysis of this kind could profitably
be extended to other aspects of the judiciary of science.

In the next sections of this chapter I shall deal, albeit in superficial fashion, with those appointed to the scientific bench by government. To whom
does government turn in search of advice? Does involvement in policy

19   H. Zuckerman and R. K. Merton, "Patterns of Evaluation in Science: Functions
of the Referee System," *Minerva*, 9, No. 1 (1971), 66–100.

making come about as a reward for scientific achievement? If so in theory, what particularistic factors appear to operate in selection of advisors? First, however, I want to say a little about an older system of advice-getting retaining some of its value in many countries today, which has important implications for stratification of the scientific community. Before the growth of the now common special advisory systems, it was usual for government to turn for scientific advice to the honorific academies or societies of their countries: the National Academy of Sciences, the Royal Society, and so on. In many countries these bodies, made up of national scientific elites, retain special links with government, remaining important sources of advice—whether formally or informally. Because of these links, well known to scientists, the elite is brought en bloc into an institutionalized relationship with government.

WHO GIVES THE ADVICE?

Elite Academies

In days when governments lacked the developed advisory systems which they enjoy today, and also lacked their own research capabilities, bodies such as the Royal Society and the National Academy of Sciences provided convenient sources of both advice and experiment. Ministers could feel sure that the desired experiments or tests would be carried out by the most competent scientists in the land. This was what Colbert and the French government had in mind establishing the Academie des Sciences in the seventeenth century. When NAS was founded in 1863, the American government did not feel the need for performance of research, but it did feel the need for a competent source of scientific advice, notably in connection with conduct of the Civil War. Although there seems to have been some feeling that such an unashamedly elitist institution was undemocratic and hence un-American, the scientific community was basically responsive to the idea of such an honorific academy.[20] Establishment of the academy was chiefly a government initiative, supported by a widely held view within the scientific community. Thus, from the outset it had the twin functions of rewarding achievement by election to membership and providing advice (if necessary, based upon experiment) on issues submitted to it by the government. Members of the academy were required to take the same oath of allegiance as were senators.

The stimulus of a later war, that of 1914–1918, brought about formation of what is today the second arm of the academy, the National Research

20 F. W. True (ed.), *History of the First Half-Century of the NAS 1863–1913* (Washington, D.C.: NAS, 1913), p. 15.

Council (NRC) (the institution's proper title is "National Academy of Sciences–National Research Council"). After tripartite talks involving the federal government, the academy, and the American Association for the Advancement of Science, NRC was established by the academy as the best means for mobilizing the nation's scientific talent in the national interest. The council was established by executive order of President Wilson "having for its essential purpose the promotion of scientific research and of the application and dissemination of scientific knowledge for the benefit of the national strength and well-being." [21] Today the National Research Council is the "operating arm" of the academy. Although it has no laboratories of its own, it is involved in collection and analysis of scientific and technical data, sponsorship of publications, and administration of substantial funds assigned for research projects and fellowships. It employs no less than 700 full-time and 4,000 voluntary consultants, with an annual budget of $19 million.[22] By virtue of its foundation through Act of Congress, and of its management of NRC, the Academy of Sciences is a "quasi-official" agency. Today the precise relationship between the academy and the council—and, in particular, the nature and extent of control exercised by the former over the latter's activities—is a subject of heated internal debate.[23] But the fact remains that the academy is enjoined to provide scientific advice for the executive, overseeing the activities of the (operational) National Research Council.

In recent years NAS has attempted, to a limited degree, to take the initiative in the science policy area. Believing the time would come when scientists would have to unite and argue convincingly for the resources which then appeared unlimited, the academy created its Committee for Science and Public Policy (COSPUP) in 1962. George Kistiakowsky, then science advisor to President Eisenhower, seems to have been the moving spirit behind its formation.[24] Its major activity, it was envisaged, would be assessment of the various fields of science: their likely substantive developments and the resources required to permit proper development. Kistiakowsky considered that, through such a process of objective compilation and submission of simple facts, scientists would be best able to argue the case for continued federal support. "I thought it would be terribly bad if we tried to get support through outright lobbying. I would not be a part of it. I thought we

21   R. S. Bates, *Scientific Societies in the United States* (New York: Columbia University Press, 1958), p. 135.

22   "National Science Policies—United States of America" (Paris: OECD, 1968).

23   J. Walsh, "National Research Council (II): Answering the Right Questions," *Science,* 172 (1971), 353. See also R. C. Lewontin, "Why I Resigned from the National Academy of Sciences," *Science for the People,* 3, No. 4 (1971), 6–8.

24   Greenberg, *Politics of American Science,* pp. 207–210.

should do it through eloquence—eloquence based on fact." [25] COSPUP has continued to work along the lines laid down by Kistiakowsky, recently under the leadership of Harvey Brooks of Harvard. One of its earliest reports was an inquiry into the problems and prospects of chemistry in the United States: a vast in-depth study funded jointly by the National Science Foundation and the American Chemical Society and monitored by a COSPUP group under the chairmanship of Professor Frank Westheimer of Harvard. The study was published in 1965, a powerful manifesto on the economic importance of chemistry and an equally powerful indictment of federal support.[26] As a consequence of the informal agreement reached between Kistiakowsky and President Kennedy's advisor (Wiesner), the federal science advisory apparatus was bound to review COSPUP studies officially.[27] Publication of the Westheimer Report (and later studies dealing similarly with other fields) was guaranteed to place the executive, in some measure, on the defensive, and Greenberg attributes later generosity toward chemistry to the impact made by this report.

COSPUP also works with and through Congress. It has prepared reports for the House Science and Astronautics Committee (or its Research and Development Sub-Committee), two of the most famous being that on "Basic Research and National Goals" and a study of technology assessment. It is possible that today NAS exerts greater influence through the links between COSPUP and Congress, on the one hand, and the wider community, on the other, than through its own direct links with the executive. Thus, writing at the end of 1970, Daniel Greenberg (an acute observer) wrote, "If budgetary growth and output of unread reports were a measure of power, the NAS would indeed fulfill its own fantasies of influence. But though more elephantine and paper-productive than ever, it is as influential with Washington as it is with the city council of Tashkent." [28] But others have publicly criticized the academy for passivity in accepting the role of technical advisor to the federal government, without divulging its views on wider issues. Among critics of the academy has been former Secretary of the Interior Stewart Udall, who accused it of being a mere "puppet" of the executive.[29] These two views are by no means incompatible and, if the ties of NAS

25 Kistiakowsky, quoted in Greenberg, *Politics of American Science* (Penguin ed.), p. 208.

26 *Chemistry: Opportunities and Needs,* Westheimer Report (Washington, D.C. NAS-NRC, 1965).

27 Greenberg, *Politics of American Science.*

28 D .S. Greenberg, "Science Under Nixon: Influence Has Declined in National Affairs," *Science,* 169 (1970), 1056–1057. (Copyright 1970 by the American Association for the Advancement of Science.)

29 "Academy Critic Shoots Wildly but Hits Sore Spot," *Nature,* 229 (1971), 151.

with government *appear* to greatly outweigh its manifestations of independence, the *symbolic* importance of the situation may outweigh its practical importance.

In Britain the Royal Society is an elite association corresponding in honorific terms to the National Academy of Sciences. Like the latter, RS has links with central government, but they are as different from those found in the United States as one would expect from the two countries' very different styles of government. The society is largely self-financing, except that it receives an annual grant from the government to disburse as it sees fit in aid of research. (This grant dates back to the nineteenth century.) Because Royal Society influence is exerted through less formal and exposed channels than is the case in the United States, its influence is difficult to assess. Representatives of the society are frequently invited to sit upon government committees concerned with scientific issues. The president of the society sits ex-officio upon the Council for Scientific Policy and is consulted on proposed appointments to this and other advisory committees. The Royal Society has rarely exerted its influence openly; it has rarely made public its critiques of policies or proposed policies. It has no equivalent of COSPUP and no links (at the formal level) with the legislature. It appears, therefore, that the society has acted as passive purveyor of technical advice, rather than as critic. But any student of British institutions is aware of the difficulties and inaccuracies attendant upon assessment of influence in terms of merely formal links.

It is untrue to say that the society has not on occasion dissented in private from proposed policies. Such dissensions may reflect the self-interest of the society, an institution with its own research, fellowship, and other activities to guard. Or its dissensions may derive from its self-image as representative of British science, and be in line with what the society's officers consider best for science. I use the word "officers" precisely, for all members of the society are neither equally committed to it as an institution nor equally consulted on society dealings with government. Thus, it is not unknown for one member to say of another (probably of longer standing), "He's much more of a Royal man than I am." It would seem that either members become socialized into the values of the institution, or commitment reflects some other (and perhaps rather interesting) variable.

At the same time, the lack of internal consultation has been pointed out by Vig. The Royal Society's view is usually that of an elite within an elite: its evidence to a government committee of inquiry into organization of civil science (in 1965) was kept secret from the general membership.[30] In times and places where no formal machinery for consultation with the scientific

30 N. J. Vig, *Science and Technology in British Politics* (Oxford: Pergamon Press, 1968), p. 183, note 15.

community exists, the Royal Society may be informally invited to fill the void. It has, on occasion, sought to put its advisory function on a more formal footing. At the outbreak of World War II, its president (Sir W. Bragg) attempted to secure for the society a formally recognized advisory and coordinatory role in the governance of science.[31] Mr. Churchill preferred, however, to rely on his personal advisor and friend Frederick Lindemann, and the society had to content itself with a very much more limited brief "to advise the government on scientific problems referred to it . . . and to bring to the notice of the government promising new developments of importance to the war effort." [32] And the society's ties with the government have remained informal, discrete, ubiquitous.

The situation in the Soviet Union is, hardly surprisingly, very different.[33] Yet, what is now the Academy of Sciences of the USSR may be discussed in a similar conceptual framework to that used for the American and British institutions. Like NAS, the (then) Imperial Academy of Sciences was founded (in 1725) as a source of both honor for scientists and advice for Peter the Great's government. In addition, it had executive responsibility for performance of research and control of the nation's institutions of both research and higher and secondary education. Through the nineteenth century it took on the aspect of similar Western bodies, and this equivalence was aided by the academy's relinquishment of its responsibilities for education in 1836. Dragged into the world of national technical need by the exigencies of the 1914–1918 war, the academy appears to have been ignored by the early Bolshevik governments. In 1927 it was given a new soviet-type constitution and attached to the Council of People's Commissars. In 1929 the academy and its staff (of over one thousand) were purged, and through the 1930s it became increasingly an organ of government. Moreover, at that time it seems to have been denied the right, characteristic of prestige scientific societies, of electing members autonomously and principally on the basis of scientific achievement. In 1935 it was formally subordinated to the Council of Ministers, which had replaced the Council of People's Commissars. The academy is tied much more closely to the government of the Soviet Union than is the case in western countries. It has

31   H. Rose and S. Rose, *Science and Society* (London: Penguin Books, 1970), pp. 69–70.

32   *Ibid.*

33   This section is based upon A. Vuchinich, *The Soviet Academy of Sciences* (Stanford: Stanford University Press, 1956); OECD, *Science Policy in the USSR* (Paris: 1969); R. Rockingham Gill, "Problems of Decision Making in Soviet Science Policy," *Minerva,* 5, No. 2 (1967), 198–208; J. Turkevitch, "Decision Making in Soviet Science Policy," *Minerva, 5,* No. 3 (1967), 430–431. See also Loren R. Graham, *The Soviet Academy of Sciences and the Communist Party* (Princeton: Princeton University Press, 1967).

been a major force in national decision making, partly because of the interlocking membership between its leadership and other important bodies (see below). Its executive responsibilities have included a virtual hegemony in planning and management of national programs of basic research (with the exception of a period between 1963 and 1965). In 1967 the academy ran 215 research establishments, employing nearly 30,000 research workers.

The main directing organs of the academy are the General Assembly, consisting of all 586 full and corresponding members, and the Praesidium. Although, in theory, the former is the supreme decision-making body, its twice-yearly meetings do not give it the possibility of exerting real power: its main task is election of the Praesidium. The latter includes representatives of the fifteen subject divisions of the academy, of the branch (or republican) academies, and includes the crucial offices of president, vice-presidents, and academic secretary. Vucinich sees in the assembly the "voice of science," in the Praesidium the "voice of government." Decisions of the government are channeled through the Praesidium to the subject divisions controlling the work of the institutes. Ideological control, the prerogative of the Communist party, is thought to be exerted through the office of academic secretary, as well as through the formal party group within both Praesidium and assembly. In addition, Turkevitch has pointed out how many chief officers of the academy (at the time of writing, academicians Keldysh, Pejve, Kirillin, Aleksandrov, and Paton) are also members of the Central Committee of the Communist party of the Soviet Union.[34]

Thus, the Academy of Sciences of the Soviet Union may be seen to combine its honorific function (in the exercise of which it boasts a greater degree of autonomy than in the 1930s) with important roles in planning and execution of research. In addition, both formal and informal links to the Communist party may be observed, a consequence of the need to ensure application of ideological as well as scientific and technocratic criteria.

One final example may be cited, demonstrating government- "academy" relations at the opposite pole from those found in the USSR. In 1949 the forces of occupation created the Science Council of Japan, in theory on the model of NAS-NRC, and with the same dual function. Two hundred and ten members are selected from the ranks of teachers and research workers in a formula designed to reward eminence and strike a balance between disciplines (applied, social, and pure sciences as well as the humanities are included) and regions. But, in spite of its representativeness and the model upon which it is based, the Science Council of Japan does not enjoy the same links with the government as other national academies. The reason for this may inhere in the political culture of the country. Over the past two

34   Turkevitch, "Decision Making."

decades, it has been suggested,[35] the sympathy of Japanese academics has been consistently with the liberal-democratic opposition. Thus the scientific establishment has not cared to identify itself with the conservative government and, even though individual scientists have worked closely with the government, "The Science Council of Japan has been cast in the role of critic." This situation is rather unique.

With the exception of Japan, elite scientific academies in most countries are closely tied to government, the result partly of the academy's current interest-seeking and political expediency, partly of a legacy of what has gone before. These ties, where they exist, are visible within the scientific community and if, as in the United Kingdom, dissension is private, the identity is all the more clear. The symbolic nature of this relationship may be more important than its instrumental nature, as I shall suggest later. But today the academies have been replaced as the major source of scientific advice by complex networks of advisory committees established by and within the bureaucracy. Though the advisory committees may be filled largely from the eminent ranks of NAS or Royal Society, members are not there as representatives of the academies. What they do represent, and how they come to be there, are complex and relevant issues.

Choice of Committee Members

I have referred to the hierarchic nature of the advisory apparatus, in which committees at the "bottom" allocate research funds to specific projects within a field, those at the "top" being concerned more with questions of broad policy, strategic choice, and so on. I have also referred to the way in which criteria appropriate to the decision process at each level must change. Within a specific field of science, one project may be compared with another essentially in terms of "internal" scientific criteria: originality, feasibility, and so on. At higher levels, a committee may be faced with a choice between two disciplines of science or between two mission-oriented laboratories. The decision process cannot be conducted in purely scientific terms: criteria to take into account may include intellectual pervasiveness of the field; likelihood of a major breakthrough; potential economic, educational, and social benefits; and so on. And, though the first kind of decision making is the form of questioning fundamental to the operation of the scientific system (tactical and strategic choices made by the researcher in the course of his work, combined with evaluation of the applicant's capability in the light of his previous performance), the second kind is very different.

---

35 Dixon Long, "Policy and Politics in Japanese Science: the Persistence of a Tradition," *Minerva*, 7, No. 3 (1969), 426–453.

And, at the highest levels of the structure, where committees may be called upon to advise upon allocations between research-performing departments of state, or upon the appropriate share of national resources for science, decision making is even further removed from what is natural in science. The simple fact of having been, or being, a researcher provides little experience for the scientific advisor functioning at these elevated levels. The scientific content of decisions is still further reduced: criteria must be essentially "external" in nature (social, economic, and so on). Effectively to represent the interests of science at these levels requires that committeemen be well versed in the political arts and fully aware of arguments carrying weight with their masters in the political climate obtaining at the time.

Such characteristic differences between the ends of this committee hierarchy serve to determine not only the work of the committees, but the status of their members and the methods by which they are chosen. When the work of the committee is within a discipline, members are united by a body of beliefs (conceptual, methodological, etc.) in the light of which they carry out their tasks. The area of consensus corresponds to the area within which they carry out their committee work and, as I suggested, the evaluative procedures are not dissimilar from those of paper referees, appointments boards, and so on. But, as one ascends the hierarchy, the area of consensus shrinks in relation to the decision area. In scientific terms, a chemist and a mathematician are unlikely to hold in common as many professional beliefs and attitudes as are two chemists. Higher level committees will usually include representatives of more disciplines than those working at the bottom. But differences are increased, the area of probable consensus decreased, by the need to take account of such factors as external intellectual ramifications, educational and economic benefits, and other external criteria. Since the scientist's views on such issues will not be a consequence of his professional career, or at least only indirectly so, they must follow largely from his extraprofessional experiences. The probability of consensus over such matters cannot be assumed to result from similarity in academic and professional life histories. A final complicating factor follows from the question of "representativeness." A scientist sitting upon a committee in which his colleagues practice a variety of academic disciplines, involved with allocation of funds to these disciplines, presumably in the light of some attempt at comparing their potentials, *may* feel he represents his own discipline. He may not know whether he was chosen as his field's "representative" or not, and his view of his role may or may not correspond to reality (that is, to his masters' view).

If those who select the scientists are anxious to ensure advice which is both authoritative and informed, on the one hand, and unanimous, on the other, these factors will influence their selection. It seems to follow that, while scientific eminence is both a necessary and sufficient condition for appointment to a low-level committee working within (and largely for) a single discipline, the same is not true at higher levels. Here we may recognize that eminence in science is necessary but no longer sufficient. Scientific eminence alone (for preference reflected in enduring and ascribed status) may guarantee advice endowed with authoritativeness and is thereby politically useful. It cannot guarantee consensus over issues entailing a large degree of extrascientific decision making, nor can it guarantee that the advice will be in any degree acceptable. No politician *wants* his scientists to advise a course of action utterly at variance with his own preferences—quite the reverse. At the most strategic levels, one may expect him to choose scientists with whom he shares—or thinks he might share—a common view on a range of extrascientific but pertinent issues. But there is no reason to expect these political preferences to percolate down to lower levels of the advisory hierarchy. Here, to reiterate, service comes about as a result of eminence and achievement within a field, and scientists accept it as an appropriate reward for achievement. As with other forms of recognition (see Chapter 3), however, it is clear that particularistic factors operate. When the reward in question is service upon a government committee, it is not unreasonable to expect either that these particularistic factors have a special salience or that they include some reference to political views.

In the United States, it appears that political acceptability has been relevant to selection of members of even single-discipline grant panels, at least over the last twenty years. Here we may see in depth the operation of one non-universalistic factor in allocation of rewards.

In mid-1969 *Science* carried out an inquiry into the Department of Health, Education and Welfare's system of vetting and appointing scientific advisors. It was clear that biomedical scientists regarded service on one of the HEW-NIH panels or committees as an important symbol of professional recognition: "If you haven't been asked to be on one of these groups, it looks like you haven't really made it in your field," one scientist noted in an interview.[36] The *Science* staff had noted a degree of suspicion among scientists that the vetting and appointment system was secretive and involved criteria other than scientific merit. It became apparent that security checks were applied to nominees to even simple grant-awarding panels. The nominating procedure allowed current members to nominate colleagues whom they thought could make useful contributions to the panel, but all federal records were then checked to see if there was any derogatory information on the nominated individual. This was done in advance of any approach

36 Bryce Nelson, "Scientists Increasingly Protest HEW Investigation of Advisers," *Science*, 166 (1969), 1499–1504.

to the individual. In July 1968, it emerged, representatives of eight bio-medical professional organizations had written to the secretary of the HEW department in the following terms:

> A number of eminent scientists have been rejected for appointment to Advisory Councils, study sections, and review committees on the grounds of loyalty or suitability. Because such rejections may be based upon irrelevant or archaic considerations, it is possible that your Department has been denied the talents of men well qualified to serve while the scientists have lost both professional recognition and the opportunity to help their country.[37]

It was alleged that both procedures and criteria were hangovers from McCarthy days; that security procedures were applied more rigorously than by the Department of Defense, even for nonsensitive grant-giving committees; and that the rejected nominee, even if he knew of his nomination, was given no inkling of the facts adduced in evidence against him. A "black-list" was said to exist, and at least one Nobel laureate (Salvador Luria) was apparently on this, together with a number of other eminent scientists. At first, the department denied the existence of such a list, claiming that a nominated individual was checked without reference to his previous clearance. Pressure was brought to bear upon a defensive department. In late 1969 the National Academy of Sciences polled its entire membership, passing a resolution declaring

> The members of the National Academy of Sciences affirm that the selection of consultants and members of advisory committees or panels by government agencies for scientific areas not pertaining to the national defense should be guided solely by the criteria of scientific competence, integrity, and judgment. An allegation of questionable loyalty should not by itself be grounds for adverse administrative action in the selection of an otherwise qualified individual as consultant or panel member for scientific evaluation of unclassified research or development.[38]

The secretary of the Department of Health, Education and Welfare responded to these repeated demands. On September 20, 1970, new procedures were officially approved.[39] These made the heads of the constituent agencies (National Institutes of Health, National Institute of Mental Health,

---

37 *Ibid.* (Copyright 1969 by the American Association for the Advancement of Science.)

38 National Academy of Sciences–National Research Council "News Report" (January 1970).

39 Philip M. Boffey, "HEW Blacklists: New Security Procedures Officially Adopted," *Science,* 170 (1970), 142–144.

Office of Education, etc.) responsible for checking the professional competence of nominees by reference to colleagues, professional bodies, and so on.
There were to be no "ex ante" security checks: nominees would be deemed
unsuitable only if their personal qualities might affect their work for the
agency (e.g., drunkenness!). It was suggested that a nominee would not be
barred "simply because he was controversial and thus might indirectly cause
the department problems in its relations with Congress." [40] Postappointment
security checks would remain (this is apparently usual even with "blameless"
agencies such as NSF), but individuals would have the opportunity to contest evidence which might lead to termination of their appointments. At any
rate, after initial appointment on scientific grounds, the scientist would have
received his professional recognition.

Few such insights into the process of selecting members of advisory
committees are available. The NAS statement above is indicative of the
scientific community's attitude toward the process, prescribed by the norms
of scientific conduct. It is perhaps scarcely realistic at the highest levels, and
at the lower levels a degree of political acceptability seems one among a
number of non-universalistic factors operating (institutional affiliation and
sex being others) in selection of panel members. These factors, particularly
his political views, may assume increasing importance in an individual's
ascendancy of the hierarchy of committees. For appointment to high-level
committees is less a reward for performing effectively in scientific research
than for performing well on committees lower in the hierarchy. This is
recognized in the scientific community, although rarely spelled out. But
widespread surprise at appointment of Edward David, less experienced in
the committee system than his predecessors, to the post of president's science
advisor was indication of implicit acceptance of the situation. The process
seems designed to ensure that those involved most directly with politicians
and senior officials have been socialized into the ways and means of the
bureaucracy and will not obstruct the consensual decision-making system.
Criteria unrelated to committee performance but relevant to the later stages
of "career development" for the statesman of science are likely to be political.
The New Critics of science object to constraints limiting the pool from
which advisors are drawn—political conformity, scientific elitism, and so on.
Normatively speaking, they are right to do so. A probably wider feeling
among younger scientists is resentment at the "gerontocratic" leadership of
science.

There are fundamental contradictions between the scientific community's view of committee service as a reward for achievement and the
political realities of the advisory role. To the individual scientist, service in
an advisory capacity is a desirable indication of his effective role perform-

---

40   HEW spokesman, quoted in Boffey, "HEW Blacklists."

ance; quite properly, he resents being denied on grounds which should not enter into the procedure for allocating recognition. At the same time many of these tasks, these advisory roles, are regarded (by both incumbents and constituents) as in a sense representative of some disciplinary constituency. A substantial segment of such a constituency may resent being "represented" by a colleague in whose selection they had no hand, and according to criteria of selection of which they are ignorant. Yet, ironically, prominent among these "secret" criteria is implicitly political effectiveness: the undemocratically chosen representative has been selected for his political effectiveness! As Sayre has pointed out, to maintain any real autonomy when on the inside means banishment from the inner councils.[41] Only to the extent that his social and political views are congruent with those of his political masters, or that he is prepared to compromise, can the representative of a discipline function effectively from the inside. This is most apparent when one looks at the very "visible" relations between the American presidents and their science advisors, but the same obtains elsewhere. To do well for his scientific constituents, the advisor needs the rapport with his master that Wiesner and Kennedy apparently enjoyed. Not for the first time, we encounter a fundamental conflict between instrumental and symbolic needs of scientists. Those who would replace their representatives by others chosen either on pure scientific merit or in light of the electorate's political views, rather than those of the government, must choose between moral (or symbolic) satisfaction and instrumental satisfaction (adequate research funds).

The Advice Givers

Here we shall look in summary fashion at the men who are chosen. How do the universalistic and particularistic criteria involved in the selection process work out when applied and reapplied in selection of those who reach the top of the tree? In practice, does scientific eminence operate as an important constraint? Are these national elites cohesive by virtue of anything more than (presumed) eminence? Do they share similar backgrounds to any extent? (How relevant are the background and affiliative factors shown by Crane and others to operate in allocation of rewards?) At the moment we can offer only a preliminary sketch of a crucial area where the sociology and politics of science meet.

One commentator on science in Washington has written,[42] "At the present time, government science advising might best be described as a sort

---

41   W. S. Sayre, "Scientists and American Science Policy," *Science,* 133 (1961), 859.

42   Meg Greenfield, "Science Goes to Washington," *Science,* 142 (October 18, 1963), 361. (Copyright 1963 by the American Association for the Advancement of Science.)

of Harvard–MIT–Bell Telephone–Caltech situation with lines out to a few Eastern universities, and to Paolo Alto, Berkeley, and the Rand Corporation." Was this ever true? Consider the President's Science Advisory Committee. Writing in 1964, after PSAC had existed in its then current form for some six years, Carl William Fischer was able to conclude regarding the committee and its substantial body of consultants that it "can fairly be said to exhibit a very high degree of homogeneity and a more or less common scientific and vocational background." [43] Of the forty-one scientists who had belonged to PSAC at the time, 70 percent had received their graduate training at seven universities. Twenty-eight of the forty-one had been academic scientists, 76 percent of whom had been associated with eight universities. In 1964, 13/16 were academics, with 61 percent attached to but three schools. There was also substantial disciplinary homogeneity: the vast majority were physical scientists, 19 percent of the forty-one being life scientists and none social scientists. Physical scientists appeared to remain on the committee for longer than did life scientists. A good deal of the day-to-day work of PSAC was done by its panels, which drew upon a substantial pool of expert consultants from a much wider range of disciplines. But the institutional bias remains: 70 percent of the academic panel members worked in the same eight universities as did 76 percent of the committee's membership.

The British science advisory apparatus displays a similar degree of homogeneity. Table 6–2 shows the constitution of two high-level advisory committees over a substantial period of time. The first, existing from 1947 to 1964, was the Advisory Council for Scientific Policy, which advised the government on the whole field of scientific and technological policy. Its chairmen were successively Sir Henry Tizard and Professor Alexander (later Professor Lord) Todd. It was succeeded in 1964 by the Council for Scientific Policy which, with a more restricted brief, reported to the Secretary of State for Education and Science prior to its dissolution in July 1972. Its chairmen have been Professor Sir Harrie Massey and (subsequently) Professor Sir Frederick Dainton.

A number of points of interest are apparent in this table. The average age of advisors has remained constant, even though that of the scientific population as a whole has fallen (today 80 percent of those with science degrees are below 50). At least as astonishing is that the percentage of the advisory group with the fellowship of the Royal Society has increased (even though some who recently sat on the council may yet receive their fellowships; most of their predecessors are at least retired). The general monopoly

43 C. W. Fischer, "Scientists and Statesmen," in S. Lakoff (ed.), *Knowledge and Power* (New York: Free Press, 1966).

TABLE 6–2
British Science Advisors 1947–1971

Period	% Aca- demic	% FRS	Average Age	INSTITUTION	Distribution of Current Affili- ation (Aca- demics)	of Ph.D.	of B.Sc.
1947–1955 (ACSP)	40	74	52	Cambridge	38	56	31
				Cambridge + Oxford	63	70	54
				Cambridge, Oxford + London	63	100	69
				All U.K. Uni- versities	= 100	100	100
1950–1963 (ACSP)	50	80	53	Cambridge	23	42	29
				Cambridge + Oxford	38	58	33
				Cambridge, Oxford + London	62	75	52
				All U.K. Uni- versities	= 100	100	100
1964–1971 (CSP)	68	85	53	Cambridge	13	46	50
				Cambridge + Oxford	22	55	54
				Cambridge, Oxford + London	61	79	73
				All U.K. uni- versities	= 100	100	100

SOURCE: Data were collected from *Who's Who* and *Who's Who of British Scientists*.
notes i. ACSP included a number of ex officio representatives of government scientific and nonscientific agencies. These were omitted: only scientists there in a personal capacity were included.
  ii. Two social scientists who sat on CSP between 1964 and 1971 were excluded.
  iii. All ages were estimated at the midpoint of the 7-year period indicated, without reference to when within that period the individual was appointed.
  iv. In the final three columns, percentages relate to those who took their bachelors' degree at a U.K. university (extreme right); took a Ph.D. at a British university (many did not have Ph.D.s, especially in the early period); were affiliated with a university at the time of their appointment (leftmost column).

of the Oxford-Cambridge-London triangle is remarkable, and has been maintained at the levels of first and doctoral degrees and of current affiliation.[44] There has been no democratization. Cohesion is greatest at the Ph.D. level: over three-fourths of all advisors have taken their Ph.D.s (if they have them) at Oxford, Cambridge, or one of the colleges of London University. Even more astonishing is that about half of those with Ph.D.s obtained them at the University of Cambridge. (The three universities accounted for 33 percent of professors, 28 percent of all academic staff in 1967.) Little significance may be attached to the changing proportion working in universities, since this involves deliberate choice and cannot be said to reflect selectors' underlying inclinations. Moreover, the increase in this percentage between 1956–1963 and 1964–1971 followed from the changing brief of the committee. The balance between areas of specialization is also an artificially arranged one and has not been included. The conclusion is that—in terms of age, scientific eminence, and educational (especially doctoral) background—scientific advisors are a remarkably cohesive elite, constantly renewed in their own image. The identity of background noted in the U.S. and British cases obtains in France too. Of ten non-medical graduate members of the Comité Consulatif à la Recherche Scientifique et Technique in 1969–1970, seven were graduates of two institutions (the École Normale Supérieure and the École Polytechnique).

All this is not to suggest that these background status factors are *explicitly* operative in the selection process (no more than in the reward system generally). The points to be made are these. In the first place, these groups are reasonably described as "elites." In the second place, in attempting to secure advice representative in terms of certain variables thought important (e.g., disciplinary balance), governments may be led into appointing advisory committees far from representative in terms of other variables. In this context, it is interesting to note that the membership of PSAC was broadened *geographically* in later years perhaps reflecting the loyalties and concerns of recent presidents.

---

44 The importance of this triangle at the research level was pointed out by Joseph Ben-David (*Fundamental Research and the Universities* Paris; OECD, 1968). Much greater importance has been attached to it by Halsey and Trow, who have tried to show that these institutions (particularly Oxford and Cambridge) have exercised both physical attraction and moral influence over successive generations of academics. Perhaps here we have the reason! See A. H. Halsey and M. Trow, *The British Academics* (London: Faber, 1971). Thus, even though there is a relationship between departmental prestige and the quality of the scientists' work (W. O. Hagstrom, "Inputs, Outputs, and the Prestige of University Science Departments," *Sociology of Education,* 44 [1971], 375–397), scientists at Oxford or Cambridge *are favored* in allocation of rewards. This is shown statistically for a sample of chemists in S. S. Blume and Ruth Sinclair, "Chemists in British Universities: A Study of the Reward System in Science," *American Sociological Review* 38, No. 1 (1973), 126–138.

The Leaders of Science

On issues on which they are unconstrained by electoral promise or political expediency, governments may simply want advice from their scientific advisors: on how best to allocate a fixed sum of money among the various fields of science, for example. To some extent, authority in such issues is delegated to representatives of the scientific community for, if allocations are portioned by a body of eminent and respected scientists, justice (in the eyes of scientists) may be both done and seen to be done. And if the externalities of such an activity are slight, the main interest of decision makers may focus on ensuring that the scientific community is reasonably satisfied. That advisors doing the dividing up should be both eminent and representative (in at least disciplinary terms) is calculated to secure acceptability of the result in the scientific community. Another major decision process in "policy for science" is determination of the appropriate level of overall financing to be allocated to science or to a major scientific agency. Governments may require the scientific community, through its representatives, to present the best case they can for whatever increase in level of expenditure for research the scientists feel that they can reasonably propose. Governments want the best possible case that they can get for a high level of increase, both to preempt any subsequent arguments from scientists-at-large and to see how the case matches up to cost-saving arguments of Treasury or budgetary officials. From the give-and-take between these two groups may emerge an appropriate level of expenditure. The government requires that its advisors demonstrate a high-level understanding of the economic and social implications of science, as well as that they retain their constituents' trust. For their job will be to explain the results of the decision process, whether formally or through their own informal contacts, and to satisfy the scientific community that the process has been equitable. Of course, the scientific community will be anxious that its own best interests are served and will also have high expectations of the spokesmen of science (though it may object on normative grounds to the methods of their choice).

In areas of policy to which science *contributes* (health, say, or defense), the government's need for scientific advice may have another dimension. Again, there will be a need for presentation of the case for contribution of research to achievement of the organizational goal: there might be an additional need for *weighty* advice to be deployed by government in countering the possibly conflicting claims of other constituency interests. In other words, scientists may suggest a greater place for research in the achievement of a given health objective than would most medical doctors; a greater place in the achievement of given defense objective than the Joint Chiefs of Staff. In general, a government may find it more difficult to resist the advocacy of a single interest group than to place competing claims in the balance together.

In a situation of conflicting advice the politician may be able to accept that case best fitting his own inclinations, and yet with the guarantee of informed, powerful backing. So in such situations the politician again requires that his scientific advisors receive the backing of their constituents so that, weighed in the balance, so to speak, they are bolstered by the full support of the scientific community. (This, in turn, may depend upon the standing of the scientific community, vis-à-vis other interest groups, with the electorate-at-large.) Thus, the influence and credibility of members of scientific advisory groups may be seen to depend upon the extent to which they visibly possess the respect of the scientific community (for which scientific eminence may no longer be an adequate guarantee); to which they possess a measure of political dexterity (perhaps acquired by long service lower in the hierarchy); and to which (together) they are representative of the scientific community.

Since committees cannot function on a day-to-day basis, especially if composed of scientists with manifold responsibilities outside their government service, their responsibilities are often partially entrusted to, shall we say, their chairmen. Presidents and prime ministers may need an individual to consult more frequently and less formally than a fifteen-man committee. For such reasons, posts have been created such as that of the President's Special Assistant for Science and Technology; in countries where there is no single post of that kind, the chairman of an advisory committee may exercise such *de facto* responsibilities. Such a man must possess the qualities delineated above to a remarkable degree, more so when he operates substantially in the public view—as in Washington—than when he operates in private—as the then Sir Solly Zuckerman (science advisor to the cabinet) in London. But for such a man, in a sense representing the whole community of science in his dealings with the executive heads of government, these qualities are but a necessary and not sufficient guarantee of political effectiveness. Although the formal power of the president's science advisor (special assistant) derived from his position as politically nominated spokesman for the scientific community, it is not possible to explain the very different degrees of influence which the six presidential advisors (Killian, Kistiakowsky, Wiesner, Hornig, Dubridge, David) have wielded over their masters in these terms alone. To explain such differences, informal relations, empathy, and personality must be invoked: the best method of understanding the operation of such factors is by comparison.

To comprehend the sources of power which those at the pinnacle of the scientific hierarchy may draw upon, we can do no better than look again at that most famous of all advisory feuds in the politics of science, dramatized so well by C. P. Snow in his Godkin lectures at Harvard: [45] The "cautionary

45 C. P. Snow, *Science and Government* (Godkin lectures, delivered at Harvard in 1960) (London: Oxford University Press, 1961).

tale," as Snow calls it, of Tizard and Lindemann, permits some generalizations which may have validity outside their immediate location in time and place.

Neither Sir Henry Tizard nor Frederick Lindemann (later Lord Cherwell) have lacked in biographers and apologists, and each figures largely in the biographies of the other.[46] A common avidity for power has been recognized in each man, leading each to wish for himself the direction of wartime British science. Both could not succeed and, when thrown together into the corridors of power, their early friendship turned to violent antipathy. But though their ambitions were similar, their approaches were very different, and they drew primarily upon what must be regarded as different sources of power.

Henry Tizard was a scientific civil servant. A scientist by training, he decided at an early age that he could never be another Rutherford or another in the mold of Rutherford's famous pupils (including Blackett, Chadwick, and Cockroft—all Nobel laureates—and Peter Kapitza). Election to the Royal Society did not disabuse him of this view. Thus in 1920, at the age of thirty-five, he entered the government service in a post involved with administration of research. He published only twenty-two scientific papers, the last in 1927—but his most famous work was to come much later. Although he left government service for the rectorship of the Imperial College of Science and Technology, he continued to occupy an advisory role within the Air Ministry, in which his service was regarded as indispensable. Tizard was an accomplished committeeman, content to work within the framework which the British system of government imposes upon its officials. One of his most remarkable gifts, highly esteemed in official British circles, was his ability to create consensus, particularly between performers and users of research. This was especially the case in the field of his major work, defense research, in which he created a situation of mutual trust between scientists and representatives of the armed forces which otherwise may never have come about. Much of the success of the Tizard mission to Washington, which inaugurated wartime Anglo-American scientific cooperation (and brought to the United State the cavity magnetron upon which radar was based, described as "the most valuable cargo ever brought to our shores"),[47] was due to its cordial

46 Apart from Snow, see Ronald W. Clark, *Tizard* (London: Methuen, 1965); P. M. S. Blackett, "Tizard and the Science of War," *Nature,* 185 (1960), 647–653, reprinted in his *Studies of War* (Edinburgh: Oliver & Boyd, 1962), pp. 101–119; R. V. Jones, "Scientists and Statesmen: the Example of Henry Tizard," *Minerva,* 4, No. 2 (1966), 202–214; J. G. Crowther, "H. T. Tizard," *Statesmen of Science;* R. Harrod, *The Prof* (London: Macmillan, 1959); Earl of Birkenhead, *The Prof in Two Worlds* (London: Collins, 1961).

47 So described by the official historian of the U.S. wartime research effort, J. P. Baxter, *Scientists Against Time* (Boston: Little, Brown, 1946).

mixture of scientists and soldiers. Tizard fitted well into the British bureaucracy, a system considering each of its officials replaceable—the job goes on, not the man. Tizard was of that breed, always recommending his own replacement.

Frederick Lindemann was very different. He was born into a wealthy German family, although his mother was American: cosmopolitan, wealthy, at ease in (and much enjoying) "high society." Success came naturally to him. Like Tizard, at an early age he was elected to the Royal Society. He remained in research much longer, occupying the Chair of Physics at Oxford, secured for him by Tizard in time of friendship, and from which he built up the Clarendon laboratory as an institution of some renown. But he was not a *great* scientist: although he published some fifty scientific papers between 1910 and 1949, he was regarded more as a remarkable intellect than as a remarkable scientist. He was eclectic in his interests, according to Snow, "an amateur among professionals which, by the way, was how the leaders of physics, such as Rutherford, always regarded him." [48] So he too opted out of the race for scientific preeminence, but more gradually than Tizard had done. In 1921, at the home of the Duke of Westminster, he met Winston Churchill for the first time. From this encounter developed a staunch friendship, shored up by enormous mutual respect, which survived until Lindemann's death in 1957, providing the key to his later influence.

While Tizard's influence developed out of his proven ability in scientific administration and policy making, out of the respect which scientists and military men alike felt for him, Lindemann's developed out of his association with Winston Churchill. A similar difference appears in their approach to the tasks which both wished to carry out. In 1936, thanks to Churchill's influence even when out of office, Lindemann was co-opted onto the Committee for the Scientific Survey of Air Defence, of which Tizard was chairman. This committee was to provide a forum for many of their disagreements, both over matters of substantive policy and deriving from personal dislike. The first of their major disagreements, occupying much of the committee's time after Lindemann's co-option, centered on the priority to be attached to development of a (then unproven) radar device. Tizard was persuaded of the feasibility and value of radar; Lindemann wanted top priority reserved for schemes of his own (infrared detection, aerial mines). Patrick Blackett (now Lord Blackett), a member of the committee, has written that "on one occasion Lindemann became so fierce with Tizard that the secretaries had to be sent out of the committee room so as to keep the squabble as private as possible." [49] But equally fundamental was a difference in *style,* reflected in their different evaluations of the work of committees such as that on which

---

48 Snow, *Science and Government,* p. 20.
49 Blackett, "Tizard," pp. 105–106.

they sat, which offered advice at a subcabinet level. Lindemann, with Churchill's support, had been agitating for a more powerful intradepartmental committee: Tizard's committee was, in his view, ineffectual and at the mercy of officials. Lindemann thought that a defense research committee, to be effective, must be not so much advisory as have executive authority over research relevant to defense research. Tizard, responding, wrote to him (July 1936), "Remember, too, that we are advisory, not executive. You may feel that this is wrong; but I do not feel that we can possibly be executive. *In any case it is not for us to decide."* [50] Lindemann could not accept such a view. Unlike Tizard, he was prepared to use political methods to achieve objectives involving reconstruction of advisory machinery. Churchill's star was then in eclipse, and in 1937 Lindemann himself stood for Parliament, albeit unsuccessfully. His grounds for so doing, according to one biographer, had been at issue in the committee, "his knowledge and experience of aeronautical matters, and his extreme uneasiness at the state of our aerial defences." [51] Tizard could never have adopted such a tack.

When, in May 1940, Churchill became prime minister, Lindemann moved with him to Downing Street. He became, first unofficially, and later (with the government office of Paymaster General) officially, scientific advisor to the prime minister. Churchill's long-standing dependence upon him for scientific and technical advice was institutionalized by his absorption into the political arm of government. Lindemann's authority derived from his daily contact with the prime minister, who continued to turn to him for briefs on scientific and technical matters. Lindemann was Churchill's sole source of such advice for much of the war. The second round of the Tizard-Lindemann dispute, staged at this time, went to the latter. Lindemann was a passionate advocate of "strategic bombing" (bombing of housing areas). Tizard and Blackett considered his estimates of the potential destruction of such bombing as five to six times too high. On their calculations, strategic bombing was a waste of resources (as later proved true). But the cabinet was persuaded by Lindemann.

In no sense could Lindemann, in office, be regarded as representative of the scientific community. Indeed, eminent members of the latter resented and were concerned about his monopoly of advice. Nobel-prize-winning physiologist A. V. Hill and others prepared a secret memorandum ("On the Making of Technical Decisions by H. M. Government") criticizing this monopoly as well as Lindemann's isolation from the scientific community: [52]

50   Letter from Tizard to Lindemann, May 7, 1936, quoted by Clark, *Tizard,* p. 141. My italics.

51   Birkenhead, *Two Worlds,* p. 151.

52   Quoted by Clark, *Tizard,* pp. 243–245. Frederick Lindemann (now Lord Cherwell)'s isolation from all but a few protégés among the scientific community has been

It is unfortunate that Professor Lindemann, whose advise appears to be taken by the Cabinet in such matters, is completely out of touch with his scientific colleagues. He does not consult with them, he refuses to co-operate or discuss matters with them, and it is the considered opinion, based on long experience, of a number of the most responsible and experienced among them that his judgment is too often unsound. They feel indeed that his methods and his influence are dangerous. He has no special knowledge of many of the matters in which he takes a hand. He is gifted in explanation to the non-technical person: the expert, however, realizes that his judgment is often gravely at fault. Most serious of all is the fact that he is unable to take criticism or to discuss matters frankly and easily with those who are intellectually and technically at least his equals.

Lindemann's unscientific secretiveness, partly reflecting his personality, had isolated him from the scientific community. His power base was peculiar to him. Tizard held the trust of the scientific community, and of the military too, for Tizard could not have worked alone. With Churchill's postwar fall in 1945, the new prime minister, Clement Attlee, invited him back to Whitehall from his "retirement" in Oxford, to serve as chairman of two newly created bodies—the Advisory Council for Scientific Policy, and the Defence Research Policy Committee. Many years later, Blackett wrote, "Widespread acclaim from both scientists and military greeted the formation of these bodies and the choice of Tizard to head them." [53] Tizard's checkered fortunes, his repeated returns from semioblivion (culminating in his appointment as, effectively, chief science advisor), combined with his complete lack of political guile (perhaps even a deficiency of political judgment), suggest that his position resulted largely from the scientific elite's unwavering support. His later appointment as foreign secretary of the Royal Society was indicative of both his status within the elite and the elite's admiration for his diplomatic talents.

Yet, in a relatively closed political system, such as that of the United Kingdom, in which crucial decisions are often made in private, there are limitations upon the political value of a power base of this kind. The support of the scientific constituency did not win Churchill's ear for Tizard: Lindemann's personality won this. That the political system implemented Lindemann's victory during the crucial days of the war has been pointed out by Snow: [54]

---

referred to by his biographer: "The other scientists in the Committee were friends and had social contacts with each other. With Prof they had none. He lived in a different social sphere, and apart from scientific arguments, had no communion with them at all." Birkenhead, *Two Worlds*, p. 192.

53 Blackett, *Tizard*, p. 112.

54 Snow, *Science and Government*, p. 56.

During the whole of his conflicts with Lindemann, Tizard had no larger body of support to call on. If he had been able to submit the bombing controversy to the Fellows of the Royal Society, or the general population of professional scientists, Lindemann would not have lasted a week.

We may therefore conclude, not surprisingly, that scientific advisors may draw power over policy either from backing of the scientific community or personal influence. The relative value of these different kinds of power base will vary as a function of the political system and the political climate. There is a sense also in which, from the chief executive's point of view, the most appropriate type of science advisor (in the sense above) may depend upon the job to be done. To the extent that the advisor's main task involves allocation of funds to basic research or formulation of policy for the organization of government research, there is need for a man trusted and respected by the scientific community. The science advisor—let us say the president's science advisor—must act as chief spokesman *for* the scientific community— only a respected scientist can do this. At the same time, he must represent the government (his political masters) and their policies *to* the scientific community. Jerome Wiesner, President Kennedy's science advisor, is on record as believing that the second of these obligations is more important [55] and, no doubt, its importance is magnified in times of financial stringency for science. Some observers of the Washington scene feel that Lee DuBridge (who resigned as science advisor to President Nixon) was unable to effect reconciliation of these two responsibilities, adopting a role first of "special pleader" for science, then changing course and trying to justify restrictive funding policies to the scientists. "The result," in Daniel Greenberg's words, "was a *tour de force* in the sequential alienation of two conflicting constituencies." [56] "Policy for science" entails the interest and attention of the scientific community: the performance of the science advisor was at the focus of that attention.

DuBridge's situation in the White House was in sharp contrast to Wiesner's. Skolnikoff, who worked in the White House Office of Science and Technology (OST, which provided the science advisor with his staff), has described the "explosion in the scale and scope of its (OST's) activities" accompanying Kennedy and Wiesner.[57] The long-standing friendship of the two men is well known, as is the fact that it led to a situation in which, in Greenberg's words once more, "the White House science adviser could stroll into JFK's office to mumble through teeth-clenched pipe that perhaps an-

---

55   J. Wiesner, quoted in "Grumbles at DuBridge," *Nature*, 227 (1970), 769.

56   D. S. Greenberg, "Washington Scene," *New Scientist*, October 9, 1970.

57   Eugene B. Skolnikoff, *Science and American Foreign Policy* (Cambridge: MIT Press, 1967), chap. 11.

other 200 million dollars might be tacked on to the budget of the National Science Foundation." [58] Skolnikoff detailed many other developments which he attributed to this special relationship: among them, strengthening of the science policy machinery in many executive departments of state (Commerce, Interior, Agriculture, etc.). The science advisor, able to draw upon his personal influence with the president, became "more aggressive." Apparently, Wiesner's successors did not enjoy the same intimacy or the same kind of positive *personal* power and influence, though there is no reason to suppose that their initial standing with the scientific community was any lower. Donald Hornig, for example, though appointed by Kennedy to succeed Wiesner, was principally associated with the presidency of Lyndon Johnson, by whom he was confirmed in office. The reduced influence of the office during this period may be attributed to two factors. In the first place, the two men appear to have felt little personal rapport. In the second place, the political situation (Vietnam), unresolved by the administration, appears to have compromised Hornig's standing with both scientists and president. The scientists felt that his apparent complicity in administration policy discredited him as a suitable voice for their opposition. At the same time, as the then Bureau of the Budget's science expert William Carey has put it, "When a President is having a lot of trouble with the intellectual community, it has a way of introducing a degree of corrosion into relations between the science adviser and the President." [59] Unlike the scientists, the president regarded Hornig as representative of the intellectual community which was giving him such a rough time.

If Johnson was not opposed to large-scale science and technology (and, apart from the Appollo ventures, the Office of Urban Technology was established by his administration), there has been a feeling that President Nixon is. This at least is the implication to be drawn from the evidence of many eminent scientists to the Daddario committee (the House Sub-committee on Research and Development, then chaired by Rep. Emilio Q. Daddario) in September 1970. Edward U. Condon, for example, opined that "the President is unaware of the importance of science in the world today." [60] Of course Nixon, DuBridge, and David are working in a situation of unparalleled disillusionment with science extending far beyond America, which has not been so much the effect as the cause of reduced funding. But the falling levels of resources, as well as direct ideological concern, are increasing the demands made by scientists on their representatives in government. To be-

---

58   Greenberg, "Washington Scene."

59   W. Carey, quoted by D. S. Greenberg in "The Hornig Years: Did LBJ Neglect His Science Adviser?" *Science,* 166 (1969), 453. A view borne out by President Nixon's abolition of the post, of the OST, and of PSAC early in 1973.

60   E. U. Condon, quoted in "Grumbles at DuBridge," *Nature,* 227 (1970), 769.

lieve that answers to all their grievances are obtainable by pressurizing these representatives is both to show a lack of understanding of political feasibility and also an unforgiveable arrogance—a neglect of public opinion. DuBridge's fall was initiated by his attempt to give voice to these misplaced demands. The scientific community must recognize the importance of public opinion and feel prepared to confront it directly. Perhaps what an advisor must do is precisely what DuBridge's successor attempted.

Edward David, a surprise appointment, came into office with little prior socialization into the ways of the Washington advisory apparatus. DuBridge had lost the president's ear and seems increasingly to have been excluded from the inner councils. But David appears to have arrived with the blessing of the president's personal staff, rather than thrust up from and by the scientific elite. He entered into a situation in which it is more difficult than ever to serve both president and a scientific constituency, for their interests are now too far apart. David seemed to have thrown himself behind the president, thereby winning his way back into those councils from which his predecessor was excluded. He seemed, for example, to have won the trust of budgetary officials in the White House. One has suggested that this is reaping rewards: "The easiest thing to do with these advisory groups is to ignore them, especially if their advice is predictable or special pleading. . . . Ed David . . . is less obviously a spokesman for a particular constituency than has been the case with OST in the past . . . . he has an audience and is much more likely to influence basic decisions." [61] By these means, by virtue of an appreciation of the need to justify scientific expenditure in extrinsic terms, acquired from a career in industrial rather than academic research, David seemed to be winning back for the scientific community some lost resources. Although he could not at the same time have functioned as an effective channel of ideological disaffection for a constituency which he did not consider that he represented, the abolition of his office in early 1973 was a considerable blow to the scientific community.

STRATIFICATION AND REPRESENTATION IN SCIENCE

In an earlier chapter, I suggested that the operation of the reward system in science produces stratification, on the basis of professional recognition received. An elite is created as a consequence both of preferential rewarding of the already successful and of the fact that, once acquired, rank and status in science are never lost: status becomes "ascribed." It is useful to review briefly the factors upon which access to the elite is founded. Quite

61 Quoted in Nicholas Wade, "Nixon's New Economic Policy: Hints of a Resurgence for R&D," *Science,* 173 (1971), 794, August 27, 1971. (Copyright 1971 by the American Association for the Advancement of Science.)

properly, scientific performance is fundamental, but I have shown that many values and prejudices common in society-at-large filter into the processes of scientific evaluation and advancement. Thus, we have noted an implicit prejudice against women in science and the advantages attaching to a prestigious academic background (explicable in terms of "sponsorship") and a prestigious current affiliation. Robert Merton has argued that the resulting stratified system is functional for science, since the elite will serve to direct scientists' attention to the more promising, exciting areas of research. Such an argument depends upon a number of assumptions about the elite; one is that accession should be solely on the basis of merit, with merit a sufficient as well as necessary condition. This may not quite be the case, and the functionalism of the elite for the intellectual progress of science must at least be questionable.

In this chapter I have focused on one form of scientific recognition. There is strong evidence that people in the field *do* regard appointment to government advisory and grant-awarding committees as a proper reward for achievement. I have discussed the criteria relevant for appointment of scientific advisors, drawing attention to the significance of political considerations. It is important, however, to emphasize that stratification in the scientific community is not due to operation of these political factors alone: They are *among* the nonuniversalistic criteria relevant to advancement in the scientific community.

The criteria determining appointment to advisory committees reflect the requirements of government: some relate to individuals' standing in the scientific community, others to extraneous statuses. Among the first group of criteria are requirements of a high level of scientific achievement (measured in practice by the possession of overt symbols of eminence) in the individual and disciplinary representativeness in the group (to the extent that numbers permit). These requirements derive from the politician's hope that his advice should not only be informed, but should receive the implied backing of the scientific community as a whole and, as a consequence, be acceptable to that community. Among the external criteria used in appointment, of increasing importance as one ascends the hierarchy is political acceptability. At the highest levels political factors play an important part in the selection process, but the work is after all highly political. It is not unreasonable that a president or prime minister should want a personal science advisor whose views are roughly in accord with his own: such an accord has proved rather crucial to establishment of an effective working relationship. However, though such procedures may constitute the best possible guarantee of the instrumental satisfaction of the scientific community, I suggested that it may often result in symbolic (ideological) dissatisfaction, both as a result of the process of selection (nonuniversalistic allocation of reward) and of the lack of ideological representation of the majority view. And we must agree that as one descends to the lower levels of the advisory hierarchy, at which political content of the job is much reduced—and at the limit scarcely present at all—the appropriateness of political evaluative criteria falls too—ultimately to zero.

In conclusion, I would like to raise two issues. The first concerns the general relations between the elite and the scientific community as a whole. In the eyes of the community, the elite must become politicized by virtue of its association with government. This follows from both the close ties frequently binding national academies of science (or the ruling cliques of such bodies) to the executive, and from the fact that the elite provides a pool from which advisors are drawn. In the limit, the elite might be seen by the more disaffected scientists as composed solely of government servants. Moreover, to the advisory elite is delegated a substantial power over direction of the scientific enterprise. In addition to the power over appointments, access to publication media, and so on which "naturally" accrue to it, the elite is delegated much power which government derives from its control of the purse strings. The functionalism of stratification for scientific progress cannot be determined solely from consideration of the provision of intellectual leadership (leadership by example), since it fails to take note of the coercive power which the elite possesses. (It can be argued, in addition, that the problem of stratification in science cannot be dealt with adequately on any kind of a functionalist model. The ideological disaffection which the scientific community may derive from association of the elite with an unpopular government may be more important to scientists than the instrumental satisfaction deriving from the success of the elite in securing a plenitude of research funds.)

This brings me to my second and final point. One aspect of the relations between elite and community hinges upon the concept of representation. One function of the scientific advisor is to represent the community of science to government; on occasion, this may create the major source of disaffection within the scientific community. The advisory elite is unlikely to be representative of the scientific community in its views on relevant issues of public policy. Because of the way it is selected, it is likely to be influenced in its perspectives by its greater age and its prestige affiliations and background, and is likely to share many views with its politcal masters. Moreover, because of its tacit association with government policies (even those with which it has not been involved), the gulf may appear wider to the average scientist than is actually the case. Yet appearances may be more important than a concealed reality. Apparent disagreements on matters external to performance of science may affect the elite's ability to exercise leadership within the scientific community even to the extent to which it is intellectually capable of, and properly should, do so.

Representativeness, to be meaningful and politically efficacious, must be based upon recognition of those dimensions across which members of the community are ranged which *they* regard as most crucial. If scientists regard their own internal differences as most significantly based upon disciplinary loyalties, representativeness involves separate representation of each discipline. This has always been the usual interpretation of the term governing bureaucratic practice. But if the scientific community were self-consciously split in other ways, and if these differences were considered more important than those deriving from disciplinary loyalties, this interpretation would break down. The idea that geographic and institutional loyalties might be important to scientists has gained ground in U.S. official circles in recent years, and the geographic representativeness of bodies such as PSAC was indeed broadened.[62] But today the scientific community seems to be dividing in new ways, based upon age and upon beliefs with respect to the social responsibility of scientists. If governments wish to preserve the concept of representativeness as some guarantee of the credibility and acceptability of their advice, they must take account of these salient new differences and involve younger scientists and those holding different views on the place of science in society. But, for their part, scientists will have to appreciate that such redefined representativeness will reduce the instrumental effectiveness of those appointed, who will be neither so versed in political in- (as distinct from out-) fighting nor so able to establish rapport with those in power.

The internal workings of the scientific reward system, the politics of pure science, and the standing of science in society are interrelated. The diminished standing of science in the eyes of the public reduces the political value of scientists' advice as a counterweight to other conflicting advice. At the same time, scientific insiders find themselves decreasingly able to appear —and, indeed, to be—responsible both to their political masters and to the scientific community, as unpopular decisions are taken. Their political impotence invisible to most scientists, members of the elite remaining on the inside become publicly associated with the unpopular policies. To the extent that such institutions as the National Academy of Sciences and the Royal Society appear to be involved with government, the scientific elite as a whole becomes publicly associated with government policies—only substantial dissension can dispel such a semblance of complicity.

This has important implications for the relations between elite and majority. The elite becomes separated from the community-at-large not only by its enduring possession (and selective receipt) of rewards for scientific work, but, in addition, by its real or tacit association with government. Thus the needs of government, and their satisfaction by selection of advisors wholly from an elite which they help to create, as well as the secret work

---

62 For one example of geographic split in the U.S. scientific community, see Greenberg, *Politics of American Science*, chap. 10.

delegated to the advisors, *politicizes* the scientific community. When the government of the day is unpopular, for whatever reason, this association of the elite acquires growing symbolic and practical importance. The internal structuring of the scientific community becomes more a matter of politics than of achievement. Only the *clear* dissension of substantial sections of the elite can inhibit emergence of this pathological situation, or halt its spread. Thus the large-scale association of people like Kistiakowsky, Wiesner, Herbert York, and Marvin Goldberger with the dissenting Federation of American Scientists may involve more *de*politicization of science than the opposite.

chapter 7
Science for the Citizen

THE ISSUES

The appeal of a participative form of democracy is powerful today, espoused alike by liberals in politics, by radical thinkers in their books, and by radical activists in the cities of the world. It is true that a measure of consensus over ends does not extend to means: many prescriptions are offered for attaining the democratic society. But one substantial strand of agreement may be discerned: the need to involve the citizenry in decision making in some way or other.

Such views are widespread in recent writings upon "science and society," and authors will frequently argue that technology has run away with itself and with politicians: all too often, technological feasibility implies technological necessity. In the *political* arena, the only case which seems to hold up against a proposed new technology is that demonstrating its technical impossibility. Writers upon such themes have emphasized society's powerlessness in the face of these technological imperatives. "When did the community decide that it wanted to invest money and skills into the development of heart transplants? Or supersonic airliners? Or chemical and biological warfare?"[1] Never, of course, because the community was unaware of the

1 Hilary Rose and Steven Rose, *Science and Society* (London: Penguin Books, 1970), p. 268.

implications and the options and, lacking any means of deciding, was disorganized. Yet the solutions to these sociotechnical questions "must . . . somehow reflect the demands, opinions, and ethics of citizens generally." [2]

Various forms which participation might take have been suggested. Calder [3] looks forward to the day when "electronic referenda" would permit involvement of the whole population in what he terms a "democracy of the second kind." This open forum would be less concerned with immediate problems than with the alternative longer term futures of society, conditioned as they are by alternative technological possibilities and priorities. Commoner's version of democracy was (and presumably is) dependent upon the emergence of citizens' pressure groups, such as that in the city of St. Louis with which he is associated.[4] The physicist Martin Perl, too, relies upon locally based pressure groups, and his emphasis upon local activity is "in part based on a somewhat generalized distrust of the efficiency of working through the national political apppartus." [5] Hilary and Steven Rose propose a rather different mechanism for public assessment of potential technological developments and continuing public control of relevant research and development programs. They suggest that research institutes should be directed by boards of management with substantial lay representation,[6] as with social institutions such as schools and hospitals. Such representation, they suggest, would have to take account of the differing class interests in society, and not reflect simply middle-class preferences. All these suggestions agree upon one facet of the scientist's role. Though they differ in the extent of involvement which they prescribe for him, all accept the scientist's obligation to inform the public of technological implications of his current research, and of the feasibility of various desirable and less desirable technological options within or deriving from his field of work.

Public awareness of and interest in science may be desirable on other grounds. The mobilization of public opinion is one weapon in the armory of all political pressure groups. For those groups lacking direct access to executive decision makers or doubting their capacity *directly* to arouse adequate interest in the legislature (or, outside the United States, doubting the relevance of ineffectual parliamentary interest), the mobilization of public concern can prove an important weapon. Under these circumstances, public awareness of and sensitivity to scientific issues becomes not so much an end in itself as a means to an end—the end being the achievement of desired policies. The more theatrical kinds of political activity are often explicitly

2   B. Commoner, *Science and Survival* (London: Gollancz, 1966), p. 94.

3   N. Calder, *Technopolis* (London: McGibbon & Kee, 1971), pp. 328–353.

4   Commoner, *Science and Survival,* p. 101.

5   M. Perl, "The New Critics in American Science," *New Scientist,* September 4, 1970.

6   Hilary Rose and Steven Rose, *Science and Society,* p. 270–271.

directed at the public (via the mass media) and, because of the publicity which they arouse, they are often highly effective attempts to awaken public interest. The disruptions of AAAS meetings by SESPA, discussed in an earlier chapter, are illustrative. Among the most successful proponents of such an approach in an initially technological area of policy making (which has since secured a certain autonomy) have been the ecological action groups. The symbolic nature of many of their activities is pretty apparent: parades in support of "smog-free locomotion," symbolic burial in a coffin of a Chevrolet V-8 engine,[7] the "Union Bay Life-After-Death Resurrection Park" project,[8] sit-ins, dig-ins, plant-ins, and so on. The need for publicity was spelled out in the *Sierra Club Handbook for Environment Activists,* which gives advice on organization of newsworthy activities, choice of name for activist groups, attracting the interest of editors, and so on.[9] And so effective has been this amazingly rapid mobilization of public opinion that already well-intentioned scientists are suggesting that the environmentally necessary ban on DDT is going to cause starvation and malaria in the developing world.

The current variety of idealized versions of democracy seem to cast the scientist into the role of purveyor of information and stimulator of public interest. But this is not a role of interest solely to the student of the politics of science: it is just as relevant for the sociologist. It is an essential consequence of the scientist's properly professional role that the greatest theoretical and practical importance attaches to informational activities of the scientific community. I made this point in Chapter 4: to claim the status of professionalism, the scientist must accept also the ethical responsibilities traditionally adhering to it. This notion received a stimulus from the recent activities of Ralph Nader, who has sought to awaken scientists' consciousness of their primary responsibilities to society.[10] A similar conception of responsibility is explicit in the statement of purposes of the (AAAS) American Association for the Advancement of Science drafted at Arden House in 1951. This statement, which provided the basis for subsequent activities of the association, recommitted it to increasing public understanding of science. The congruence between professional scientific and informational activity has been given added weight of a different kind by Commoner, who wrote in his best-selling book *Science and Survival*:

> At first many of us tended to regard the task of public education as something apart from our professional life, as a kind of civic responsibility for

7 C. Humphrey, "Doing Ecology Action," in *Ecotactics: the Sierra Club Handbook for Environment Activists* (New York: Pocket Books, 1970).

8 R. M. Pyle, "Union Bay: A Life After Death Plant In," in Humphrey, *Ecotactics.*

9 John Zeh, "Getting into Print," in Humphrey, *Ecotactics.*

10 See pp. 124–125.

which our general training prepared us. But, as we began to examine the available evidence for ourselves it became clear that it did not entirely support certain statements by government officials regarding, for example, the fall-out hazard.[11]

And, indeed, the work of Commoner and his associates (discussed in greater detail below), which frequently involves compilation of data to be judged against the data and conclusions of other scientists, spans the gap between the internal (communal) and external (societal) communication of scientific findings.

This increasingly relevant aspect of the professionalism of scientists finds some tangible form in the institutional structure of modern science. As briefly indicated, it forms a major part of the work of such general scientific societies as AAAS (which has near relatives across the world). Moreover, it is *increasingly* central to the work of these societies, as a consequence of both their declining utility within the scientific community as channels of specalist communication (a result of the growth of specialist societies) and the growth of professional consciousness among their members. Reasons for this, obtaining in many countries, were spelled out by a committee of review looking into the work and functions of the Australia and New Zealand Association for the Advancement of Science (ANZAAS). The committee proposed a new constitution, later adopted, for ANZAAS, setting out its aims as

> the advancement of knowledge by bringing together scientists of all types and occupations so that they may interchange information and ideas, and making the public aware of what science is, what scientists do, and of the applications and implications of scientific discoveries to their everyday affairs and those of the nation.[12]

In the United States the same ideals also find expression in the scientists' information movement, of which Professor Commoner is a leading light. The philosophy of the movement has been expressed thus:

> The scientific information provided to the public by all groups and individuals affiliated with SIPI is developed and presented by professionally competent scientists with traditional scientific accuracy and objectivity. Attention is given to divergent views; moral and political judgments are not made. Information is made freely available to all, in the belief that the dissemination of such information is necessary for a democratic society in a technological age.[13]

11   Commoner, *Science and Survival,* p. 103.
12   "New Image for ANZAAS," *Nature,* 227 (1970), 652.
13   "SIPI" is the Scientists' Institute for Public Information. The extract is from a

The primary aim of this chapter is to elaborate upon this aspect of scientific professionalism by examining the information work of such organizations and of individual scientists. In the final chapter, I shall attempt to study the scientist's role in the wider context of popular participation in decision making in this area: what use is and can be made of information he provides? The organization of the present chapter is as follows: In the next section, we shall examine the institutionalized commitment of the scientific community to increasing public understanding of science, institutionalized in and exercised through general societies such as the American and British Associations for the Advancement of Science, and through the scientists' information movement in the United States. Then we shall look at presentation of science in the mass media, for it is mainly via the media that the public becomes aware, on anything but a small scale, of scientists and scientific happenings. What are scientists' attitudes toward the popular presentation of research? What are the problems and prospects of science reporting in the press, TV, and so on? Finally, tentative conclusions will lead on to the final chapter and a discussion of participatory democracy in this area of policy making, as it presently functions.

THE INSTITUTIONAL COMMITMENT OF THE SCIENTIFIC COMMUNITY

In a recent speech Dr. William Bevan, chief executive of AAAS, said, "I have been asked to address myself to the question of roles and responsibilities of scientific organizations like the American Association for the Advancement of Science in our changing world. My thesis . . . is that the special responsibility of organizations like the AAAS is both to interpret science to the larger society and to participate in assessing its consequences."[14] So the major *institutional* commitment of the scientific community to increasing public understanding and awareness of science has been taken up, as a principal task, by organizations such as AAAS. Yet this is a role adopted gradually, over many years, and not only because of its importance, but equally because of the loss of other more prestigious tasks to the specialist scientific societies. In addition, those who lead AAAS and BAAS would be the first to admit that practice has scarcely matched up to commitment. No one who studied only their actions could deduce their high ideals. And yet the gap between ideal and action may be narrowing, at least in the United States, and, to the extent that it is, I believe this is due mainly

---

publicity leaflet. (Copyright 1970 by the Scientists' Institute for Public Information, New York, N.Y.).

14 W. Bevan, "The General Scientific Association: A Bridge to Society at Large," *Science,* 172 (1971), 349–352.

to the growing forces pulling rather than pushing (as was the case) the association toward a new conception of its proper task.

BAAS and AAAS

The origins of the British Association for the Advancement of Science are to be found in the writings of a number of eminent nineteenth-century scientists who observed what they took as a decline of science in England. Sir Humphrey Davy began a book on the subject,[15] but died before it was finished. In 1830, however, Charles Babbage published his *Reflexions on the Decline of Science in England.*[16] His thesis received strong support, among others, from Sir David Brewster (principal of St. Andrews University, later of Edinburgh University), editor of the influential *Edinburgh Review.*[17] In his review, Brewster called upon his readers to observe "the science of England . . . struggling for existence, the meek and unarmed victim of political strife." His review ended with a suggestion: "An association of our nobility, clergy, gentry, and philosophers, can alone draw the attention of the sovereign and the nation to this blot upon its fame." For a model upon which to base such an association, Brewster and his friends looked to Europe. And in the Deutscher Naturforscher Versammlung, founded by the Munich physiologist Lorenz Oken in 1822, they discovered it. A number of British scientists attended the peripatetic early annual meetings of the German association, including Babbage and Robert Brown (famous as the discoverer of Brownian motion). One visitor described its objects in an article written in the *Edinburgh Journal of Science:* principally, to promote intercourse between men of science—but thence other benefits would flow. The meetings "draw public attention to science and scientific men, and make people inquire concerning both them and their pursuits. They exalt science in general estimation, and with it those who devote themselves to its advancement; and, above all, they spur on the Governments of the different States to examine into and ameliorate the condition of their scientific institutions." [18]

This was the rationale behind the "advancement of science" move-

---

15 Sir Humphrey Davy (1778–1829) was perhaps the chief prophet of applied science of his generation, as well as a most brilliant chemist. He was a close friend of the romantic poet Coleridge and became president of the Royal Society.

16 Charles Babbage, Lucasian Professor of Mathematics at Cambridge, was an irascible and near universal genius, whose intellectual contributions extended far beyond mathematics. He also invented a remarkable calculating engine, a mechanical ancestor of the electronic computer.

17 O. J. R. Howarth, *The British Association for the Advancement of Science—A Retrospect 1831–1931* (London: British Association, 1931), chap. 1.

18 James F. W. Johnston, F.R.S., quoted in Howarth, *British Association,* p. 7.

ment in its early days: to establish the collective identity and visible strength of science, and thereby to bring pressure to bear upon the government. It was upon this model that David Brewster wished to base his "Society of British Cultivators of Science": few felt confidence in an aristocratic, lethargic Royal Society being able to drag itself out of its torpor. The role of the British Association for the Advancement of Science (as it became known), which first met in 1831, is defined by its statutes:

> To give stronger impulse and more systematic direction to scientific inquiry; to promote the intercourse of those who cultivate science in different parts of the British Commonwealth with one another and with foreign philosophers; to obtain more general attention for the objects of science and the removal of any disadvantages of a public kind which impede its progress.

In its turn, BAAS was to provide the model for an American association, and a number of prominent Americans attended its meetings, among them Agassiz, Bache, and Rogers.[19] Through the concern of these men and others, an earlier association of geologists was, in 1848, turned into an American Association for the Advancement of Science. A committee composed of Agassiz, Rogers, and Benjamin Peirce drew up its objects, with two major spheres of interest: "to promote intercourse between those who are cultivating science in different parts of the United States" and "to procure for the labours of scientific men, increased facilities and wider usefulness."

Both bodies were primarily promotional in their early days, although this always involved promotion of the "visibility" of scientific activities. Apart from annual meetings, BAAS published the results of papers communicated to it, used membership dues to support scientific research, and attempted by a variety of means to bring pressure to bear upon a government which at that time had little interest in science. Similarly, many early AAAS leaders hoped it would become a major source of advice for the government; some saw it, in addition, as a research funding agency, an instrument for the support of research by the federal government. Thus, before foundation of NAS (in 1863), of the major specialist societies (the American Chemical Society in 1876, the American Geological Society in 1888, the American Mathematical Society in 1894, the American Physical

---

19 Louis Agassiz, a naturalist, was America's foremost anti-Darwinian; Alexander Dallas Bache (Professor of Natural Philosophy at the University of Pennsylvania) was "the great tycoon of American science." Benjamin Peirce was a Harvard mathematician. These men, together with the great experimental physicist Josph Henry, were leaders of the group known as the *Lazzaroni* (scientific beggars). According to Reingold, the same group was largely responsible for foundation of the National Academy of Sciences. W. B. Rogers was a geologist who founded MIT. See Nathan Reingold, *Science in Nineteenth Century America* (London: Macmillan, 1966).

Society in 1899, etc.), and of NSF in 1950, the AAAS could reasonably regard itself as a suitable vehicle for all necessary (but then unperformed) functions within the scientific social system: communication, representation, recognition, and so on.[20]

The development of these other institutions, particularly the specialist societies, forced the general associations to seek more limited but valid roles. AAAS, as we have seen in an earlier chapter, found inadequate the efforts which it could make toward accommodation of new specialist disciplines: it was unable to prevent their separate organization. So, in retrospect, the need for a new role should have been obvious early in the twentieth century—but, whether on account of institutional inertia or because vision was befogged, it was much later before one encountered even tacit acceptance of the need for change.

In the United Kingdom, the scientists most concerned by the renewed failure of British technology, who saw its cause in the continued neglect of science by government, found BAAS inadequate as a vehicle of protest as early as 1903. In that year, the astronomer Sir Norman Lockyer, then president of the association, decried the lack of a collective voice for science, urging BAAS actively to work for wider utilization of scientific research in industry. Rebuffed, he founded the British Science Guild, which survived till 1936, when it was reabsorbed by BAAS. In the early days of the 1914–1918 war, the British Science Guild was among those institutions which successfully prompted the government of the day to establish the Department of Scientific and Industrial Research (DSIR).[21] Among other activities, this provided government support for academic research, rendering obsolescent BAAS's smaller scale activities of that kind.

It was not until the middle of the twentieth century that a conception of any new, more realistic role for the general associations emerged on either side of the Atlantic. Indeed, the British association's centenary volume of 1931 spends many pages decrying the (even then many) prophets of doom, and seems happy enough to take attendance at annual general meetings as an adequate indicator of organizational vitality.[22] But in 1946 AAAS adopted a new constitution, adding a new and more outward-looking mission:

> The objects of the American Association for the Advancement of Science are to further the work of scientists, to facilitate co-operation among them,

20   Early history based upon R. S. Bates, *Scientific Societies in the United States,* 2nd ed. (New York: Columbia University Press, 1958), pp. 73–79, 125–128.

21   I. Varcoe, "Scientists, Governments and Organized Research in Great Britain 1914–16," *Minerva,* 8, No. 2 (1970), 192–216; see also W. H. G. Armytage, *Sir Richard Gregory: His Life and Work* (London: Macmillan, 1957), pp. 67–136.

22   Howarth, *British Association,* chap. 9, "Retrospect and Prospect."

to improve the effectiveness of science in the promotion of human welfare, and to increase public understanding and appreciation of the importance and promise of the methods of science in human progress.

These new objectives—science in the promotion of human welfare and increasing public understanding of science—were given further weight by the so-called Arden House Statement of 1951, the result of concentrated deliberation on the association's future role. This statement provided the philosophy behind, if not the exact blueprint for, later AAAS activity. Finally, in 1969 a new conception of what AAAS should be was put forward, but one still in line with the growing support for an essentially mediatory role between scientists and the community. The 1969 statement committed the association to try for a vastly broadened membership and to attempt to communicate with a much wider public.[23] A more activist role was implicit in the particular emphasis now given to "the promotion of human welfare." In seeking its new, wider public, AAAS has chosen five specific groups to address most directly: the scientific community itself, youth, teachers and educational administrators, national and local government leaders, and science journalists.

In 1957 the functions of BAAS were also reviewed, and a number of recommendations were made. The programs of the annual meetings should be planned, including more joint meetings between the disciplinary sections, and with greater emphasis upon science for nonspecialists. Committees should be set up to study specific problems relating to science's impact on society, with conferences on such issues of immediate interest arranged. BAAS should take a greater interest in the young, stimulating discussion of innovation in science curricula and seeking generally to arouse youthful interest in science. The association should arrange more lectures, publish more pamphlets and a monthly journal, and persuade the press, radio, and television to devote more attention to science, particularly to science's impact on society. A measure of the influence of that committee and its proposals is that in 1970 another committee was established, which in 1971 made not dissimilar recommendations. This committee saw "a continuing need for an organization of national standing, embracing both the natural and social sciences, where the impact of science as a whole can be discussed and where problems involving a number of scientific disciplines are pursued in their social context." [24] Its role would be "to stimulate public interest in science and technology and their relation to social problems, and to promote the advancement of science and technology to the benefit of the community."

---

23 P. M. Boffey, "AAAS (III): Is Order of Magnitude Expansion a Reasonable Goal?" *Science,* 172 (1971), 656.

24 Report of Committee of Review (London: BAAS, 1971).

Its future role might comprise provision of information to the public (especially young people) about applications of science and technology; promotion of interdisciplinary discussions and studies, outside the program of annual meetings, on topics of common scientific interest and centering on the social impact of science and technology; and publishing the results of such studies.

Both associations are anxious to fill the vacuum which replaces any effective forum for the *integration* of science (a necessary complement to its disintegration on specialist lines) and to fill other roles in diffusion of the results of science and consequences of these results to a lay audience. In broad terms, these objectives are in line with the new conception of scientists' professional responsibilities. In neither case have activities matched up to objectives, disparities partially attributable to exceedingly cumbersome styles of government, and in the British case to financial strictures. But the American association is much more virile and *relevant:* a difference far greater than its seventeen-year advantage in age would indicate! BAAS boasts 4,600 members, about half students and schoolchildren. In the United States there are something like eight times as many research scientists as in Britain, yet AAAS has 130,000 members, or nearly thirty times as many. Thus, to see how their rather similar aims work in practice, we have good reason to concentrate upon the healthier of the two siblings.

A few characteristics of the membership of AAAS are pertinent.[25] Among the 130,000 members, scientists predominate (there are many more than engineers); about half the members work in universities, one-fourth in industry, and one-fourth elsewhere. Roughly equal numbers of physical and biomedical scientists constitute the membership, and something like three-fourths of the membership of NAS belongs. It is less easy to characterize the decision-making structure. The highest authority in the association is the 530-member council, composed largely of representatives of the three-hundred or so affiliated scientific societies prior to the 1972 reorganization.[26] The council was unwieldy in the extreme, met rarely, and exercised only a limited control on most issues. It did elect the board of directors (from 1973 to be elected by direct suffrage), made up of eminent scientists (although rarely the "statesmen of science" interested in assuming real power over science policy)[27] but, again, prevented by their substantial inter-

---

25 P. M. Boffey, "The AAAS (II): What It Is and What It Tries to Do," *Science,* 172 (1971), 542.

26 The "affiliation" of breakaway scientific societies was one method used to cope with the disruptive effects of specialization. The idea was embodied in a revised constitution of 1899 and in 1901 they were granted representation on the council. See Bates, *Scientific Societies,* pp. 125–126. However, from 1973 a 95-man Council will be elected directly by the membership.

27 Boffey, "The AAAS (II)."

ests from exercising much control. Finally, there is the permanent staff, headed by the executive officer—a post filled for many years until recently by Dael Wolfle. Its relatively heterogenous membership and uncertain power structure have doubtless contributed to the broad chasm between the association's objectives and its activities. Its two best-known activities are its weekly periodical *Science* (read by many more than its 163,000 subscribers) and its peripatetic annual meetings. The meetings, attended by something like 7,000 participants (and often 1,500 speakers), receive widespread coverage from the mass media. Over the past two to three years they have moved away from their earlier emphasis upon more technical papers, till now a large proportion deal with problems either of interest to a wide range of scientific disciplines or, indeed, to scientists and laymen alike. "But the process of change is far from complete, and it is meeting with resistance," [28] partly as a result of the autonomusly planned programs of the affiliated societies, which hold specialist meetings concurrently.

Even if the content of *Science* and that of the annual meetings are changing in the direction of AAAS's long-professed commitment to increasing public awareness of and benefit from science, the association is still open to the charge that it is doing little to "improve the effectiveness of science in the promotion of human welfare." This is, as we have seen, a frequent charge from radical quarters: the association's idealistic goals make the discrepancy all the greater, the charges of ineffectiveness all the more powerful. (It is not only the press coverage available which makes the meetings such a favorite target of the radical scientists, it is also the loftiness of the association's aims and its claim to be the most representative of all scientific bodies in America.) To vindicate itself of these charges, AAAS must add to its activities, since such objectives can scarcely be attained by a small circulation magazine and an annual meeting. Is it doing this? Well, its study of the use of herbicides in Vietnam was a first, and not unimpressive, excursion into "technology assessment." A follow-up is a current study of power production directed by Professor Barry Commoner.

A number of other possible activities, now under discussion, would carry the association farther in the appropriate direction. It has been suggested that, since AAAS is both independent and "representative of the scientists-in-the-street," it could give expression to views of the scientific community as a whole on matters in which NAS represents solely the views of a small elite.[29] It could provide a forum for discussion of the implications and effects of new and existing technologies. It could do more work at the regional and local levels, as by providing schoolteachers with more scientific support and allowing them to identify more closely with the scientific enter-

28 Boffey, "AAAS (II)."
29 Boffey, "AAAS (III)."

prise. One or more new publications, aimed at a lay audience (whether youth, politicians, or some other segment of society—this remains for discussion), are proposed. Formulation of a program of such specific new activities is a step toward matching actions to objectives. And that this should be happening, that new initiatives are at last being taken, is indicative of a changing feeling among the membership-at-large. May it not be that inculcation of a greater degree of understanding of science among the general public is an objective acquiring new salience for many members of the scientific community?

The scientific community has not demonstrated a high regard for external dissemination of research findings, for "popularization," in the past. No matter how much lip service has been paid by the scientist to the need for such activities, their pursuit has secured for him not the slightest professional recognition. Hagstrom has suggested that publication of popular science articles (even textbooks) may "reduce a man's prestige within science" and that "most popularizations . . . are written for money and not out of a sense of duty." [30] And certainly the scientist unwise enough to allow the mass media access to his findings before publication in the scientific journals has called upon himself the wrath of the scientific community. Samuel A. Goudsmit's editorial on the subject in *Physical Review Letters* (a prestigious "rapid publication" physics journal) has achieved a certain notoriety: "In the future we may reject papers whose main contents have been published previously in the daily press." [31] Another editor who has drawn attention to negative sanctions applicable against scientists who do not recognize their duty to inform the scientific community before the world-at-large is Franz J. Ingelfinger (of the *New England Journal of Medicine*). He has acknowledged that "if the principal ideas of an article, as well as its crucial data and most important figures, had already appeared in a medical news medium" this would count against the article in deciding upon publication.[32] These views may be presented as expressing an inevitable conflict of loyalties for the professional and Hagstrom explains his findings in terms of the scientific community's protection of its monopoly as source of professional recognition. Thus, scientific articles written for popular consumption are rewarded not in the coin of professional recognition, but in cash or popular acclaim, and they may prove so attractive to the individual that his loyalty to the scientific community is weakened.

30 W. O. Hagstrom, *The Scientific Community* (New York: Basic Books, 1965), pp. 34–35. For an alternative view, see below, p. 233.

31 Editorial, *Physical Review Letters,* January 1, 1960.

32 F. J. Ingelfinger, "Medical Literature: The Campus Without Turmoil," *Science,* 169 (1970), 831.

But from another perspective these twin loyalties are complementary rather than competitive. As the editorial remarks suggest, some scientists have preferred to allow the mass media first access to results on issues of social consequence, so that the data are available for interpretation by laymen before they have been attested by the professional community. Making results available before professional consensus has been reached on their validity poses possible dangers, as two recent cases demonstrate. The first case concerns the apparent inducement of cancer in "smoking beagles" by scientists associated with the American Cancer Society, who announced their findings to a press conference.[33] The study was described in a press release by the society as the first laboratory demonstration of cancer causation, which "should have a significant impact on the smoking of cigarettes in this country, and will probably lead to a reassessment of advertising claims and policies of the cigarette industry." Not surprisingly, the study aroused considerable interest. But its submission to the scientific community, and consequent validation of the findings leading to these claims, was held up, since publication was refused first by Ingelfinger's journal (on grounds of prior disclosure) and then by the *Journal* of AMA (on grounds of scientific flaws). Following its appearance in a third journal, both methodology and actions of the American Cancer Society were attacked by a past president of the College of American Pathology, who wrote, "There is obviously a grave risk that the public, and even the medical profession, may be misinformed and misled by (relying on prepublication results). Science can do well without the hanky-panky that results from such unwarranted publicity." [34]

A second relevant case was the "Sternglass affair," in which even those scientists most opposed to deployment of nuclear weapons demonstrated their disapproval of widespread public support for their cause based upon apparently fallacious evidence.[35] Professional responsibility would appear to require that scientific findings of substantial social importance and public interest be submitted first to the professional community for attestation. Ideally, the scientific *community* must be first to sanction release of research results to the public and so, in a sense (the temporal sense), the *first* duty of the scientific investigator must be to his professional colleagues. And it

33 "Premature Puff for Smoking Beagles," *Nature,* 230 (1971), 547.

34 *Ibid.*

35 Professor Ernest Sternglass, in a series of widely read articles in the professional scientific press, in the *Bulletin of the Atomic Scientists,* and in *Esquire* (all received widespread additional coverage by the daily press), tried to show that the explosion of nuclear devices had caused the deaths of 400,000 children either directly or via genetic damage. Though his work was powerful evidence against the testing of nuclear devices, it was widely denounced as "unscientific," by even the most committed scientists.

is the release of certified information to the public that AAAS and other organizations growing out of the increased "professionalism" of science regard as their primary role.

Even more overtly committed to such work is the "scientists' information movement" in the United States, a movement maintaining close ties with AAAS. Indeed, the association's Committee on Science in the Promotion of Human Welfare (which has done precious little) expressed the rationale behind the movement in a report issued in 1960. The committee wrote:

> We conclude that the scientific community should, on its own initiative assume an obligation to call to public attention those issues of public policy which relate to science, and to provide for the general public the facts and estimates of the effects of alternative policies which the citizen must have if he is to participate intelligently in the solution of these problems. A citizenry thus informed is, we believe, the chief assurance that science will be devoted to the promotion of human welfare.[36]

The Scientists' Information Movement

The movement dates from 1958,[37] a time when there was a widespread fear of possible dangers of atomic radiation from the frequent nuclear tests. There was a growing belief among scientists and informed nonscientists that if citizens were to prove capable of making the choices which control of the new technology demanded (for example, to balance the cost of Sr^{90} in the teeth and of I^{131} in the thyroid gland against the military benefits of atmospheric testing), they would need the requisite information. Committees of scientists emerged in a number of American cities, motivated by a desire to supply the technical information, but in nontechnical form. Before 1963 these committees were concerned almost exclusively with biological effects of radiation. By that time about twenty-three groups had sprung up, of varying sizes but all widely acknowledged as authoritative sources of information on such issues. In 1963 the limited test ban treaty was signed, and there was a decline in public anxiety—"Those committees which had not become firmly established with some financing and staff to provide continuity gradually dissolved." Those surviving "recognized the need for a national body, which would co-ordinate their work, provide a clearinghouse for information and a means of communication between the groups. Such a

---

36 "Science and Human Welfare," *Science,* 132 (1960), 1475–1476. (Copyright 1960 by the American Association for the Advancement of Science.)

37 Except where otherwise indicated, all quotations in this section are from "Scientists, Citizens, and SIPI: A Short History," *SIPI Report,* 1, No. 1 (1970), upon which the section is based. (Copyright 1970 by Scientists' Institute for Public Information, New York, N.Y.)

body would also be able to find funding for existing committees and to work with scientists interested in organizing new groups. The committees also wanted to expand their programs beyond radiation information, and looked to a national organization for direction and guidance."

As a result of this feeling, and motivated also by an anxiety over other dangers to the human organism and its environment which were becoming apparent, over one hundred scientists met in New York in February 1963 to found the Scientists' Institute for Public Information (SIPI). It was established as a nonprofit educational body, "to seek out, inform, and enlist scientists of all disciplines in public information programs." Moreover, the movement's interests have widened to include many more issues than those raised by testing of nuclear devices: not surprisingly, questions of environmental quality loom large, and SIPI groups are involved with chemical and biological warfare, and even biological and social effects of drug abuse, race, and population growth. In brief, the rationale of the movement lies in its attempt to break monopolies on information. An agency of the federal government may constitute both major source of a given form of pollution and only source of expert information: the monitoring of radiation damage is one case in point. "How many nuclear scientists can you find that aren't on the AEC payroll, in the AEC labs, or on AEC university grants? The AEC has budgetary power over most of its potential critics who have expertise. It is almost impossible to find alternative sources of information." [38]

The institute's policies are determined by its Board of Directors. The president is Margaret Mead, the chairman Barry Commoner; other members include Rene Dubos and Edward Tatum of the Rockefeller University. There is some overlap in leadership (as in ideology) with AAAS. The institute's work is carried out largely by local chapters, the best known of which is perhaps at St. Louis. The work of the St. Louis group has been described by Professor Commoner, one of its founders, in his book *Science and Survival*. It was born in April 1958 from a widespread local anxiety over rising strontium[90] levels in milk. A need for information developed as mothers began to worry about whether their children should drink less milk. "Several of us in the St. Louis scientific community decided to try out the idea that scientists could usefully inform the public about such matters. In April 1958 an organization of several hundred St. Louisians was formed. Citizens helped to define the areas of public concern." And so, "when the scientists had educated themselves they announced that they were available as public speakers on the facts of fall out." [39] So the Committee on Nuclear Information (CNI) was born. A major part of its activity has been interpretation of data on science-generated issues of public concern. Commoner quotes an

38 Scientist member of CCEI, quoted in *Science*, 168 (1970), 1324.
39 Commoner, *Science and Survival*, p. 102.

issue relating to hearings of the congressional Joint Committee on Atomic Energy on consequences of a nuclear attack, given certain assumptions. Committee witnesses gave "details" of the effects of an attack, but information available to the public from press coverage was bitty and mentioned only a few highlights. Committee on Nuclear Information scientists prepared their own analysis of the information, focusing specifically on the physical, medical, and ecological effects of the two bombs "assigned" to St. Louis. How best to present the analysis, given the desire to communicate effectively with the population-at-large? "After several unsatisfactory drafts had been prepared, they decided that the scientific reports could be translated without loss of detail or accuracy into fictional accounts by supposed survivors of the hypothetical bombing of St. Louis." [40] The account was published in the group's monthly magazine: its effectiveness was such that it was reprinted in the *Saturday Review,* in other major papers in the United States, and in six other countries. Subsequent activities of the group, now called the Committee on Environmental Information, include the following. An extensive report has been compiled on air pollution in the city, dealing with such questions as: Did the 1967 Air Quality Act prove effective in reducing air pollution? Are present air pollution standards adequate? What do citizens and officials need to know about air quality? Work is also under way on possible hazards of certain insecticides, and a report is in preparation on the effects of underground nuclear testing.

A group concerned with the implications of nuclear explosions is the Colorado Committee for Environmental Information (CCEI). This group of twenty-four scientists (twenty are physicists) was founded in 1968, operating with no staff and a minuscule budget.[41] The bulk of its work consists of producing and distributing scientific reports on specific environmental questions. In accordance with SIPI philosophy, it does not indulge in general polemics or "environmental activism." Its president, biochemist Peter Metzger, has pointed to the major impact of the organization: "Perhaps the most important change is the realization by the public in Colorado that they are no longer forced to rely on information about pollution from the polluters alone." To date, the major concerns of CCEI have been possible dangers from the store of nerve gas at the Rocky Mountain Arsenal, radioactive contamination from underground nuclear explosions, and plutonium pollution at the Rocky Flats plant of Dow Chemicals–AEC. On May 11, 1969, this plant caught fire—a conflagration involving $50 million of plant and $20 million of plutonium. In view of the potential dangers deriving from this

---

40   Commoner, *Ibid.,* p. 105.

41   "Colorado Environmentalists: Scientists Battle AEC and Army," *Science,* 168 (December 18, 1970), 1324. (Copyright 1970 by the American Association for the Advancement of Science.)

release of plutonium to the environment, the press approached CCEI for information, and the committee prepared a list of questions which it felt required official answers. Press coverage of the CCEI questions was extensive, and as a result the governor of Colorado requested that AEC send top officials to provide authoritative answers. The officials met members of the CCEI, but the scientists were dissatisfied with their responses. CCEI members felt that data supplied by the AEC did not accurately record the extent to which plutonium had escaped into the environment and, indeed, that the commission lacked adequate methods for measuring radioactivity in, for example, soil around the plant. CCEI scientists decided to carry out their own tests, obtaining data which resulted in some subsequent contention between CCEI and AEC. The ensuing public controversy was reflected in local politics, and a number of bills calling for closer monitoring of emission have been presented, with Colorado congressmen taking up the matter with AEC.

Other issues worrying SIPI groups include the dangers of lead poisoning from common materials. For example, in May 1969 the New York SIPI published a bulletin on lead poisoning in slum children, organized a conference on the same subject, and arranged a congressional briefing. Technical information was provided for the state legislature Committee on Public Health, part of which was included in a bill designed to prevent lead poisoning by prohibiting use of lead paint in certain places. Similarly, a Minneapolis city ordinance controlling the use of lead paint resulted partly from a lead poisoning screening carried out jointly by Minnesota CEI and the Biomedical Students Committee for Social Responsibility of the University of Minnesota.

SCIENCE AND THE MASS MEDIA

The majority of citizens are not in touch with local committes of SIPI, nor do they attend meetings of AAAS or read magazines such as the *Scientific American* or counterparts in other countries (*New Scientist, La Recherche*). For the majority, even in the world's most highly educated nation, the mass media represent the major, or sole, source of information on current issues in contemporary science. Not surprisingly, it appears that (in the United States, at least) an individual's education is the chief predictor of his overall scientific knowledge and, moreover, that his education largely determines the attention he will give the scientific content of the mass media. After completion of formal education, increases in scientific knowledge come mainly from the media: daily newspapers, above all, followed by TV and magazines. [42] It is possible to reach many more people through the

42 Wilbur Schramm, "Science and the Public Mind," in *Studies of Innovation and Communication to the Public* (Stanford: Stanford University Press, 1957).

mass media than directly through books, pamphlets, lectures, and so on, and those producing the latter with a didactic eye on the general public may quite rightly focus concern on attracting the attention of the former. Thus, those who interpret scientists' responsibilities in terms of "the need to inform" must, as they do, welcome the attention of the media. Consequently, though their motivations differ, AAAS and SIPI, on the one hand, and the "eco-activists" and SESPA, on the other, are united in their attempts to capture the attention of the media, and thereby a mass public.

If the "new professionalism" to which I have so often referred is gaining ground within the American scientific community (if not elsewhere), then an increasing cordiality between members of the community and the representatives of the media would seem likely. An improving situation would foreshadow not only a new status for the profession of science journalism, but new demands upon its practitioners as well. The fact is that scientists' failure to recognize their responsibilities to society has given rise to a situation characterized by something less than complete harmony. Because of the vital mediatory role which science journalists must play in any attempt to inform the public-at-large to any significant extent, a brief discussion is called for on relationships between scientists and science writers.

Scientists and Science Writers

In ninteenth-century Britain popularization of science was a major concern of the many eminent scientists instrumental in founding such organizations as the Society for the Diffusion of Useful Knowledge, the Society for the Encouragement of Arts, Manufactures, and Commerce (now the Royal Society of Arts), the city and guilds institutes, and the mechanics institutes.[43]

Whether because they saw in wider diffusion of knowledge a source of national prosperity or of individual happiness for the workingman as a consequence of better appreciating the bases of his work, many scientists devoted a good deal of energy to popular lectures, organizing educational institutions, museums, and exhibitions, including the Great Exhibition of 1851.[44] Thomas Henry Huxley was perhaps the greatest scientific popularizer.

What of popularization by scientists in modern times? We have little information on this issue: the most valuable study of which I am aware relates to France. Although its results are not strictly applicable outside the

---

43   D. S. L. Cardwell, *The Organization of Science in England* (London: Heinemann, 1957), chaps. 3, 4.

44   It is estimated that a full 17 percent of the British population visited the 1851 exhibition. In 1850 the total newspaper readership was only 2 percent of the population: daily newspapers sold only about sixty thousand copies altogether. See R. Williams, *The Long Revolution* (London: Chatto & Windus, 1961), chap. 3.

French context, they are worth reviewing, both for the suggestive nature of the analysis and as a stimulus to possible replication.

Boltanski and Maldidier carried out a survey of popularization activities of two hundred scientists working in university and CNRS (Centre Nationale de la Recherche Scientifique, a government-supported basic research agency working in close association with universities) laboratories in the Paris region.[45] Some 36 percent indicated that they had been involved in some way in "popularization of science" (defined principally by reference to medium of publication or transmission). A major issue for later analysis is that involvement is highly dependent upon academic rank: whereas only 4.5 percent of *"assistants"* (a grade roughly comparable with American instructors) have been involved, the figure rises to 33.5 percent of *"maîtres-assistants"* (assistant professors), to 50 percent of *"maîtres de conference"* (associate professors), and to 85 percent of (full) professors. Moreover, it was more usual for the less senior to *refuse* to write popular articles or speak on the radio than for senior colleagues. Other significant differences appeared between more and less senior scientists: the less senior preferred to use traditional pedagogic methods of presentation and to write about the classical experiments of scientific history, while the more senior preferred to discuss their own work; many more junior scientists tended to think their popular writings should retain the impersonal style usual in science, while a sizeable fraction of their seniors thought that the scientist should be presented with his work. The explanation of all this is the feeling that popularization must be "in the name of all science": "Scientists who popularize by this act put the whole of the scientific enterprise in the public eye. It follows first that only the most senior scientists, possessors of a sort of 'permanent pedagogic authority' may have the right and the privilege to popularize";[46] moreover, the scientific community delegates responsibility for representing it to its most senior members, most subject to its control. At the same time, the media, "themselves incapable of conferring legitimacy and authority upon the facts they disseminate and divulge," *require* the prestige of the most eminent scientists to add authority to their contents.[47] Young scientists must recognize that they do not possess the right to invoke the authority of science, and that they may be penalized in their careers if they exceed their rights.

For the scientific elite, according to Boltanski and Maldidier's analysis, the situation is very different. First, and this is an observation worthy of particular thought, this elite is oriented away from the scientific community and toward other social elites (*les autres fractions des classes dominantes*), and

45 Luc Boltanski and Pascale Maldidier, "Carrière Scientifique, Morale Scientifique, et Vulgarisation," *Social Science Information*, 9, No. 3 (1970), 99–118.

46 *Ibid.,* p. 103 (my translation).

47 Compare the needs of government, discussed in the last chapter.

toward holders of administrative, political, and economic power within these elites. Second, popularization is not a unique phenomenon, but only one among many extrascientific tasks which the elite is called upon to perform. The fact that such scientists prefer to discuss their own work, rather than to discourse upon scientific history, may then be seen as (at worst) disguised self-seeking or (at best) a political gesture in the interests of science.[48]

We have no comparable data for the United Kindom or the United States. The analysis would seem tenable, however, and we may see popularization as dysfunctional (from a career point of view) for the young scientist. Because he is not seen to have the right to speak on behalf of the scientific community any such gesture on his part will be punished. For the elite, whether or not we accept its identification with other national elites, popularization may be regarded as a political gesture—just as was the foundation of BAAS and AAAS. We must distinguish between popularization for pedagogic and for political aims.[49]

It is unlikely that American and British scientists are engaged in popularization of science to the same extent as French scientists (at least, those working around Paris). Perhaps any lesser involvement may be attributed to greater professionalization of science journalism in these countries (if this is the case). Certainly, popularization of science is substantially the province of this group today.

In the 1920s science journalism emerged as an identifiable occupational role. The doyen of British science writing, J. G. Crowther, tells in his memoirs how he approached the most famous British newspaper editor of the time (C. P. Scott of the *Manchester Guardian*), asking for the post of "science correspondent." Told there was no such occupation, he replied that he intended to invent one![50] In the United States science journalism emerged at about the same time, with the American Society for the Dissemination of Science founded in 1919 and Science Service in 1921. Indeed, the early history of science writing in the United States is virtually the history of the publishing organization Science Service. Its establishment was largely a result of the initiative of the Scripps-Howard newspaper chain's founder, E. W. Scripps.[51] Scripps' objective was to increase the amount of newspaper coverage given to science, and the new organization was designed principally to

48 It is interesting to note that scientists prefer "prestige" publications (read by influential members of society)—such as *Le Monde, Le Nouvel Observateur*—for their popular articles.

49 For analysis of the functions of popularization, see W. Ackermann and R. Dulong, "Un Nouveau Domaine de Recherche: la Diffusion des Connaissances Scientifiques," *Revue Française de Sociologie,* 12 (1971), 378–405.

50 J. G. Crowther, *Fifty Years with Science* (London: Barrie & Jenkins, 1970).

51 Based upon P. M. Boffey, "Science Service: Publishing Pioneer in Financial Trouble," *Science,* 169 (1970), 1182.

disseminate science news to the press. In the 1940s it was transmitting material to more than one hundred client periodicals, with a combined circulation of over 30 million. But growing postwar interest in scientific matters, stimulated by The Bomb and Sputnik, led to development of science writing as a proper branch of journalism. "Soon every daily paper with any pretensions had its own science writer or science staff, the major wire services and supplementary services had expanded their science coverage, and new specialized science syndicates sprung up." [52] In a sense, the objective behind Science Service had been realized, and its centrally provided press service was of diminishing importance. And, indeed, the press syndicate was abandoned in 1970. But Science Service is still of interest, both for the scope of its activities and for its formalized relationships with key scientific institutions. In particular, the latter is indicative of the formal recognition (if no more) that the scientific community has been prepared to accord the dissemination of science news. Current activities include the publication *Science News*, with circulation of 115,000 among scientists, teachers, students, and laymen; the annual review *Science News Yearbook*, aimed at students and teachers; and various other sponsored publications. It also boasts a flourishing foundation-sponsored youth program, which organizes such events as the International Science Fair and the Science Talent Search. Via its Board of Trustees, Science Service is linked to the National Academy of Sciences, the AAAS, and the National Research Council. Of the fifteen-member board, three trustees are appointed by NAS, three by NRC, and three by AAAS; the remainder are representatives of the Scripps Trust and the profession of journalism. Discussion preceding the proposed merger with AAAS suggested how similar are the ideals and activities of the two organizations.

Today in the United States, and elsewhere, science journalism is a recognizable profession or branch of a profession. In the United States the professional body, the National Association of Science Writers (NASW), was founded in 1934; by 1966 it had 254 members.[53] Krieghbaum estimates that today around three hundred and fifty to five hundred individuals in the United States, working in the media, devote half or more of their time to basic science, technology, engineering, or medicine—this being the NASW's criterion of eligibility.[54] In Britain a probable majority of science writers have a background in the discipline but not, apparently, in the United States. The typical American science writer is a college graduate in his mid-forties, likely to have majored in literature or journalism rather than in science, although probably he has taken more science courses than the typical arts graduate.

---

52   *Ibid.*

53   H. Krieghbaum, *Science and the Mass Media* (London: University of London Press, 1968), p. 86.

54   *Ibid.*

Most seem to gravitate to science journalism from journalism rather than from science.[55] Few, it seems, find their lack of specialist training a serious barrier to effective interpretation of scientific developments. One survey found that, on average, the areas in which they felt most competent were medicine, general biology, psychology, and the physicochemical sciences—at the opposite extreme, they were much less happy with the engineering sciences, agriculture, metallurgy, statistics, and mathematics.[56] The geographic distribution reflects the distribution of scientific activity in the United States. Seven areas account for three fourths of all members of NASW, based in metropolitan New York, the District of Columbia, and the states of California, Illinois, Massachusetts, Texas, and Michigan.[57] Fifteen states boasted no members at all.

The science writer's work is beset by difficulties, exacerbated by its producers' traditional disinterest in, or distaste for, the general diffusion of scientific news. The scientist, in offering a scholarly article to his peers, is relatively sure of his audience. Some scientific periodicals are read more widely and quoted more frequently than others, and scientists will seek to publish their work in these more "prestigious" journals. But this is a question of prestige, not of attracting readers, and the journalist or TV producer must win his audience from the population-at-large. (This partially accounts for the lurid headlines often attached to sober pieces of scientific reporting, which so annoy scientists.) In addition to these fundamental difficulties of interpretation, the science writer has had to face the uncooperativeness of the scientist, whether born of genuine hostility or of an unpreparedness or inability to draw the implications from his work which are potential sources of popular interest. A further source of friction lies in the media's propensity to personalize, for mass audiences are often as interested in the man as in his discovery: does he, for example, correspond to one of the familiar sterotypes ("absent-minded scientist," "Dr. Strangelove," etc.)? Scientists often object to this, for scientific norms stress communality and reject the association of discovery and discoverer; overt publicity-seeking will almost always secure the obloquy of the scientific community. However well rooted in scientific norms, it is hard to deny that this conduct embodies an abdication of professional responsibility which leads scientists to condemnation rather than cooperation. "After one programme which looked at the implications of some current cytological research, a professor of pathology whose work had been shown complained to a *Nature Times News Service* reporter that

---

55   *Ibid.,* p. 84. This may result from the greater professionalization of journalism in the United States than elsewhere.

56   W. E. Small, "Training of the Science Writer" (M. Sc. dissertation, Michigan State University, 1964), quoted by Krieghbaum, *Mass Media,* pp. 95–96.

57   Krieghbaum, *Mass Media,* p. 87.

the public had more to fear from the ability of mass communications media to distort, misrepresent and terrify than from any of the biological experiments shown in the programme. His co-worker wrote to *The Times* describing the film as an awful warning to scientists who are tempted "to come into the market place of public-relations men, mass media, fashion photographers, and pop stars!" [58]

How widespread is this antagonism? According to the ex-head of science broadcasting at BBC (British Broadcasting Corporation) TV, "There is a small but significant group of hostile scientists." [59] G. Wood has attempted to assess scientists' reactions to the media's treatment of their work.[60] His study focused on the 1955 meeting of the American Psychological Association, held in San Francisco. Five reporters, representing three local papers, *The New York Times,* and Associated Press, covered a conference involving presentation of 420 research reports, 64 symposia, and 23 speeches. Each psychologist whose presentation was covered in detail was sent a copy of the relevant article(s) and asked to comment upon the accuracy. Twenty-two out of thirty-three psychologists responded, and the work of these twenty-two had generated fifty-seven articles and stories. Six (equivalent to eight stories) felt that the whole point of their papers had been missed. Apparently, this was the case when their work *refuted* some well-known or easily remembered theory (e.g., that "there is honor among thieves"). Seventeen gave examples where the articles "overgeneralized" the data, "exaggerated" or "oversimplified." Thus, although the majority had some complaint, these grievances seem associatable, on the whole, with writers' attempts to render the technical findings intelligible or interesting, rather than with sheer incomprehension.

Scientists' attitudes toward reporting of scientific developments, whether deriving from general opinion or specific personal experience, are worth further study. How concerned are sections of the scientific community to actively ensure accurate interpretation and reporting of their work? Such attitudes, to reiterate, are closely identifiable with professional consciousness (although a much less stringent test than Nader's "whistle blowing"). This connection has recently been pin pointed with great force by molecular biologists James Shapiro and Jon Beckwith. In 1969 their important research in

58 R. W. Reid, "Television Producer and Scientist," *Nature,* 223 (1969), 455. Boltanski and Maldidier, "Carrière Scientifique," found that attitudes to science writers were more favorable among senior scientists. Frequently, junior scientists proclaimed the need of science journalists for a solid scientific education. Moreover, whereas 64 percent of the junior scientists felt that "popularization does not give an accurate image of science," this was true of only 50 percent of the senior scientists.

59 Reid, "Television."

60 G. Wood, "A Scientific Convention as a Source of Popular Information," in *Studies of Innovation and Communication to the Public.*

molecular genetics led to isolation of a pure bacterial gene, and the work was reported amid a welter of purposively induced publicity. There is little doubt that the popular press greatly exaggerated the implications of the work, seeing it as a much larger step toward the specter of "genetic engineering" than it was in fact. The researchers' deliberate encouragement of publicity and of exaggerated interpretation of their work was viewed with disfavor by the scientific establishment.[61] The young researchers' attempts to vindicate their actions are of interest. They admit, "The press greatly inflated the importance of our particular piece of work. This was due in part to some of our own statements, which were misleading. It is true, however, that progress in the field of molecular genetics in the last few years has been extraordinary. We felt that the isolation of pure *lac operon* DNA was a graphic, useful, and easily understood example of that progress. . . . We did not publicize our work in order to add to our own or Harvard's prestige or to make a plea for more money for basic research . . . we wished to make (a) political statement." [62] Perhaps *only* a "plea for more money" would have constituted an acceptable reason for descent into the world of popular communication— though it, too, would have been a political statement. Shapiro, Beckwith, and Eron go on to say, "As we see it scientists are obligated to inform the public about what is happening in their secluded fields of research so that people can demand control over decisions which profoundly affect their lives." Beckwith has since made the point that the scientist who fails to ensure that the implications of his published work are spelled out for the general public is acting unethically in the same way as the scientist who consents to work on secret research.[63] To say this is at last to put the scientist's two obligations—to his professional community and to society-at-large—in the scale together, and to find their weights equal.

Science in the Mass Media

Science enters into the media in a host of ways: this variety, combined with the obvious variations in nature and standards of the organs of mass communication, produces a complex situation. Different newspapers and programs have different interests in science: for one it is a source of edification, for another of news, for a third of horror or fantasy (what Krieghbaum calls the "Gee Whiz" style of reporting!).

A few quality newspapers and television programs present factual material about science with educational aims. Discoveries with implications for

61   See editorials, *Nature,* 224 (1969), 834, 1241.

62   J. Shapiro, L. Eron, and J. Beckwith, "More Alarums and Excursions," *Nature,* 224 (1969), 1331.

63   J. Beckwith "Discussion" in W. Fuller (ed.), *The Social Impact of Modern Biology* (London: Routledge & Kegan Paul, 1970), p. 232.

thought in general (cosmological, ontological, ethical)—such as of quasars and pulsars, influencing man's understanding of the origins of the universe; of the genetic code; of the nonconservation of parity—such discoveries in pure science are widely reported. As we have seen, in general they are not reported in the way that scientists themselves report upon results of their work, and these differences can produce tensions. Thus, the media frequently exaggerate the importance of a discovery: everything may appear as a "major breakthrough." Also, by reporting a discovery in isolation, as they must, they conceal the fabric of science. They fail to reveal the extent of the dependence of every successful piece of work upon the successes and failures of countless other scientists: the result is a "heroic" attitude toward and reflection of science out of tune with the views both of social scientists and of scientists themselves. There are exceptions, and a very few newspapers and television and radio programs seek genuinely "to supplement the study of science in . . . secondary schools by linking classroom studies with today's scientific research and programs." [64] Krieghbaum instances the Minneapolis *Morning Tribune* which has, since 1963, apparently contained a weekly page of current science for schools; it has received an AAAS citation for this. It is more usual for scientific education to comprise a by-product of news reporting with a scientific content: in general, it is the newsworthiness of a discovery, rather than its scientific or purely educational significance, which determines its coverage. An important source of science news is provided by annual meetings of scientific societies, which are "news" merely by virtue of taking place. The meetings of AAAS and BAAS receive special coverage, even though the material presented is scarcely original in scientific terms. Nevertheless, an account of work delivered at one of these annual get-togethers is news, and the associations for the advancement of science do feel a commitment to present material of wider interest. Hence, both the material selected for these meetings and the style of its presentation (hopefully) make it especially suitable.

It is interesting to compare the perspectives on science which readers of different newspapers will obtain (assuming they read the pieces in question) from their papers' coverage of such meetings. How much is covered, and how? I have compared the coverage of one meeting of BAAS by four British newspapers: such a procedure could readily be extended. The four national newspapers are *The Times* (an "establishment," center-right paper, with a 1967 circulation of 364,000); *The Guardian* (a liberal paper popular with academics, with a 1967 circulation of 281,000); the *Daily Mirror* (a left-wing mass circulation paper with a 1967 circulation of 5,282,000); and the *Daily Express* (a right-wing mass circulation paper with a 1967 circulation of 3,948,000). Their coverage of the British association's 1968 meeting, held in September of that year, is represented in the table on page 240.

64 Krieghbaum, *Mass Media.*

TABLE 7–1

Reporting of the 1968 BAAS Meeting

Headlines

Express	*Mirror*	*Times*	*Guardian*
		Irresponsible Science Could Quench Human Life	Scientists' Responsibility for Evil Results
Pact to Ease Britain's Brain Drain Losses Urged	Pouring Down the Brain Drain—A Multi Million Cash Loss	US Urged to Take First Step in Ending "Brain Drain"	US Ultimate Beneficiary of 'Human Capital'
	TV Travel Show Made Viewers Stay at Home	Study of Television's Impact	Assessment of TV Impact on Viewers Often Widely Wrong
			Daily Noise That Deafens Without Being Much Noticed
			Goods in the Pipeline by 1978
			Sales Related to Union Militancy
		New British Potatoes Are Needed	New Purpose Comes to Farming
			Principal Backs Loans for Students
			The Nose Is Still Best for Smelling
		Taking the Smell from the Factory	'Clean Works'
		Danger in Antibiotic Spread	Danger from Farm Antibiotics
	That Old Tribal Fetish—a Family Saloon	Car Seen as Tribal Symbol	Use of Computers in Psychology
		Robots as Secretaries Soon	Chatterbox

Express	Mirror	Times	Guardian
			From Fast Trains to Cars you Hang on the Wall
Deep freeze Banks on Way for Surgery	'No Hope' for Deep Freeze Cure	Small Hope of Life after Freezing	Frozen Bodies Will Stay Dead
	Big Hunt for the Guinea Pig Children	Scottish Children Read Better	
	Time for a Bite —by a Mosquito		
			The Sombre Shadow over Science
			(Leader): Science and Civilization
	Monkeys Became Junkies to Help War on Drugs	Power of Cannabis 'a Fraud'	Drug Addiction 'Has Become an Epidemic'
	The Boys Who Say: We Steal	One Boy in Three Says He Stole	The Boys Who Steal Even from Relative
			Paymaster of Mental Arithmetic
		Doubts Arise on the Big Bang Universe Theory	Oddities of Space Hold Shocks for Future
			The Proper Study of Computers
			Why Women Get Depressed
	Giant Jets That Will Seat 1000		1000-Seat Aircraft Foreseen
	Mr. K. and the Revisionist Climate	Dry Climate and Mr Khrushchev	
		Streams a Key to Diet	

TABLE 7–1 (continued)

Express	Mirror	Times	Guardian
		Automation Has Its Limits	
It's Safer to Fly with the Big Boys	'TV Heroes' Plan for SOS Squads	Plans for Nuclear Rescue	The Robens Plan for National Rescue Service
The Sun Is Going to Blow Up, Says Professor		Day the Sun Explodes	Escape in the Solar Explosion
			No Cure Likely for Salmon Disease
		Sex Secrets for Computer	Sex Problems Ahead for Computer
			Ignorance of the Oceans
		Seaweed May Hold Key to Fallout Protection	
		School by Bus adds to Children's Worries	
	Young Scientists to Get Big Boost	Drive to Bring Youth into Membership	Young Scientists to Have their Own Organization
	Mysterious Moles Put Boffins in a Hole	Moles Are Not Invaders	Garden Problem Reduced to a Molehill
	Join the Teaset of Tomorrow —It Won't Break	Unbreakable Cup	Unbreakable Cup on the Way
	New Look at London	Mr Grimond Attacks 'Jungle North of Watford' View	Need to Vitalize Fringe Areas
	Brighter Children Prefer Good TV		Children Own Best Judges

Though this comparison of headlines is scarcely a scientific "content analysis," it is indicative of the varying styles of reportage, and shows the items chosen for inclusion in these four very different papers. Look first at the number of presentations reported by each newspaper. The serious papers (which are, in fact, much larger) have included many more items than the popular papers. This is to be expected, since their readerships will contain many more of the highly (and scientifically) educated. Comparing *The Guardian* with *The Times* and the *Mirror* with the *Express,* we find that the left-inclined papers have reported more in each case, though in socioeconomic terms their readerships will be similar (number of items were 33 and 25; 16 and 4). The styles differ between the cooler objective approach of the serious papers ("Drug addiction 'has become an epidemic' ") and the more eye-catching style favored by the popular papers to obtain their readers' attention ("Monkeys became junkies to help war on drugs"). But there are also striking similarities. In all cases, with a very few exceptions, the issues presented refer to a simple social issue or object to which all readers can relate: this may apply less to the detailed discussion, but it is true of the headings under which they are presented. Each heading involves one or more terms within the experience of everyone, almost irrespective of education: noise, car, boy, aircraft, sex, cups . . . and so on. Rarely, can the papers assume the reader's interest: they must catch his eye. The text of the article may be more dispassionate, but the scientists' ire may have been roused!

Without copies of the original scientific papers, it is not possible to check on the accuracy of the reporting: the best we can do is look for internal consistency. And we find that in only one instance is there what *may* be a mistake of interpretation, judged solely by this criterion of consistency. The *Express* reports the coming of "deep freeze banks"; the other papers report what we might take as the tone of the scientist's remarks, that this is not feasible. Of course, we do not know whether or not the paper's subeditors felt that the impossibility of such an invention was much less interesting than its possibility. It is noteworthy, however, that this was the only one of the four papers not represented by a professional science writer.[65]

The relative commitment of the four papers to science, as measured by coverage of BAAS, has been maintained in succeeding years and many tentative conclusions above would be further substantiated by repeating the same exercise. We may limit ourselves to noting just one or two points from report-

65 The "inexpert" papers make use of headlines almost guaranteed to annoy the scientist—in the attempt to interest or amuse their readerships. The "press" given to the experimental drug L-Dopa is illustrative. L-Dopa is being used to cure Parkinson's disease under laboratory conditions but in 2 percent of cases, it seems, it also functions as a sex stimulant. The "sex angle" was played up as a reason for reporting the drug by a few papers. *The New York Times* ran the headline "A Drug is Shown to Combat Parkinson's Disease"; the New York *Daily News* had "Parkinson's Disease Drug Can Spur Sex Drive" (January 15, 1970).

ing of the 1970 Durham meeting. We find all papers still picking out items which could more easily be related to the interests and experiences of their readers and, on occasion, changing the emphasis of the talk in attempting to play up this relationship. The proceedings of the association's sociological and medical sections provide cases in point. Professor Michael Banton presented a paper to the former in which he discussed the police role in a changing society. This was reported in detail by *The Times,* which pointed to his espousal of the police as a "service" rather than a "force." A problem for police, the speaker observed, was the lack of uniformity in law enforcement. *The Times* went on: "The biggest present source of inconsistency in law enforcement was that concerning offences committed on private premises. Many employers and employees seemed now to accept a certain level of theft as constituting one of the 'perquisites' of the job." [66] This point, accounting for only 5 percent of the 25 column inches given over to the talk by *The Times,* received prime emphasis in the *Mirror* and *Express* reports. The *Mirror's* heading was "Pilfering Is Just a 'Perk,' " whereas the *Express* took up an example given in illustration as the theme of its heading: "Sacked . . . Security Man Who Tried to End Stealing." No mention of the police!

On September 5, 1970, *The Guardian* reported papers given to eight different sections of the association: a total, including headlines, of 107 column inches. Of these, 2½ column inches were devoted to a paper delivered to the "general" section on "the scientific community"—its values, norms and so on—by the physicist-professor John Ziman, and 10 column inches to a biomedical paper on a demonstration of the effect of physique upon stress by Dr. P. K. Bridges. Intriguingly, the mass circulation *Mirror* took Ziman's article as its lead on the day's procedings: "The 'Supermen' Scientists: It's Just a Lot of Rubbish, Says Professor." Bridges' presentation was the second feature reported. *The Guardian,* under a heading "Physique and Stress Related," had described tests on volunteer students seeking to measure their reactions to a situation of psychological stress (in fact, a mock examination!). The author's conclusions, quoted verbatim, were that "the findings tentatively suggest that some factors related to body build are of importance in the patterns of physiological response to a purely psychological stress." Conclusions had suggested greater *reaction* (in terms of increased heartbeat, hormone production, etc.) on the part of the thin than of the muscular. Press reaction was intriguing. The *Mirror,* the *Express,* and *The Times* all opted for personalized headings. *The Times* had "Big Men Best in Crisis"; the Express, "Hey Fatty, You've Got a Nerve—But Skinny is Getting the Jitters"—both emphasizing the advantages of reduced response. The *Mirror,* on the other hand, took a different line—"The Skinny Action Men"!

66   *Times,* September 8, 1970, p. 3.

These comparisons are not intended to suggest widespread incompetence in reporting science: far from it. I have shown, I hope, that simple misunderstanding is relatively rare, and that the situation is much more complicated than that. The two factors underlying most criticisms of science reporting by the media are difficulties of interpretation, when scientific papers have a definite, not sufficiently spelled out public interest, and reemphasis, to draw attention to some relatively minor point which achieves greater popular than scientific interest.

Postwar years have seen a great expansion in the reporting of science news, an expansion almost paralleling the growth of science itself. A 1951 survey found that something like 60 percent of daily papers in the United States had doubled their allocations of space to science within the previous decade. Similar surveys carried out in 1958 (one year after Sputnik) and 1965 found that the growth was being maintained. The indications in 1958 were that 38 percent had doubled their allocations within the previous decade (and 37 percent had increased it by half); in 1965, 47 percent appeared to have doubled (and 30 percent increased by half). In no case had the coverage given to science in the "space age" declined.[67]

The increase in journalistic interest in science over the past years is not in "science for its own sake," and rarely is it an increased commitment of the media to pure education. The growth in interest is in science as news, and it is concentrated in those areas of science furnishing news—of whatever sort. The 1951, 1958, and 1965 surveys of editors' opinions, to which I have referred, asked respondents to indicate also those areas of science, technology, or medicine in which their (daily) papers had special interest. The results, though not surprising, are worth quoting.

Medicine and public health issues remain of great interest and the ex—science

TABLE 7–2
The Main Areas of Editorial Interest in Science

Issue	Placings and Percent of Editors Mentioning		
	1951	1958	1965
Medicine and public health	1—82%	2—57%	2—69%
Atomic energy	2—76%	3—55%	4—35%
Agricultural science	3—40%	4—32%	3—44%
Satellites and outer space	—	1—80%	1—78%
Military science	7—18%	5—28%	10—12%

SOURCE: Abridged from Krieghbaum, *Mass Media*, p. 78.

67 Surveys were carried out by N.A.S.W. and N.Y.U. School of Journalism. Reported by Krieghbaum, *Mass Media*, pp. 72–73.

editor of the New York *Herald-Tribune* suggested that, in gathering potential material, science writers should divide their time about equally between
medical and all other technoscientific sources.[68] Interest in atomic energy
declined, as it gradually became something of a commonplace—lacking (it
seemed) either human or eschatological interest! From 1957 "space," "satellites," and suchlike shot to the top of the list: here was the action. News of
launchings, manned flights, and (eventually) the moon walk were received
by a rapt public. Probably no "scientific" achievement had ever before received such public attention. Note, too, the rise of interest in "military
science" in the cold 1950s. One may reflect upon what a similar survey today
would show. Perhaps we would have found that the primacy of extrater-
restrial doings would have been maintained through 1969 but that, following upon the first moon walk, interest (like financing of the Apollo program)
then declined. Surely there can be little doubt that the state of the environment would head the list today (although this is not solely the interest of the
science writers). We may then, if we choose, reflect further upon the implications of these lists.

The scientific content of the mass media is not confined strictly to its
newsworthiness. It plays a prominent part, for example, in the advertising
which we find in many organs of mass communication. We may hypothesize
that, though popular *interest* in science determines the fraction of news space
which it attracts, the *status* of science determines (and is partly determined
by) its place in advertising. Let me elaborate. I am not referring to the use
of insights gained from the social sciences in planning advertising campaigns,
but to the verbal or visual *evocation* of science in advertisement. To be
effective, advertising must appeal to certain qualities, attitudes, or beliefs
common among the audience (prior knowledge of the audience is imperative). Appeals can focus on snobbery, religiosity, romanticism, sexuality,
implicit respect for views of high-status groups, and so on. The frequent—
or once frequent—use of science falls into this latter category: "Scientific
tests show . . ." A British campaign to persuade people to drink more milk
and not to drink when about to drive emphasized the "scientifically proven"
effect of milk on levels of blood alcohol (see illustration). In visual terms
the man in the white coat is often most persuasive; so is the swinging pointer
on some hypothetical "whiteness meter" or "sweetness meter" or "softness
meter." Firms may attempt to boost their images by showing how deeply
they are involved in research: the advertisements of the chemical company
Hoechst (United Kingdom), seen in many scientific and popular periodicals
today, are indicative of such an approach. The appeal is to the status of
science: if people are not impressed by a commitment to science or per-

68 Quoted by Kreighbaum, *Mass Media,* p. 109.

Milk helps you survive your friends' Christmas drinks

Alcohol concentration in the bloodstream.

50

40

30

20

10

WITHOUT MILK
See how alcohol taken on a virtually empty stomach *stays highly concentrated in the bloodstream,* causing dizziness and fatigue.

WITH MILK
If you drink a pint of milk first, the level of alcohol in the bloodstream *is much lower.*

1½   2   2½   3

Hours of time since experiment began.

This graph is adapted from NATURE (Vol. 212, No. 5066). It shows you how a pre-party pinta effectively reduces the level of alcohol in the bloodstream.

Result: a clearer head, a heightened sense of well-being – and a brighter morning after!

And don't forget – if you're driving, make sure it's only milk you're drinking.

It's a medical fact!

Reprinted with acknowledgment to the National Dairy Council of England and Wales.

suaded by its findings (real or imaginary), such campaigns will be ineffective.[69]

And, finally, "science, like all the other information sources, has been mined to provide human interest materials, vicarious thrills and amusing anecdotes." [70] Science may provide sheer entertainment—whether wittingly or unwittingly—especially when the debunking nonspecialist comes to it. Take, for example, the following extracts from a review of a terribly serious TV documentary on astronomy (from my own favorite newspaper!).

> Of the three million faithful who watch Horizon . . . it may be said (so I shall proceed to say it) that a million understand it, a million think they understand it, and a million get a sharp stabbing pain just here at the base of the skull.
>
> I don't hesitate to stand up and be counted with the third million, being one to whom the phrase "total disintegration of an ageing star" suggests only James Cameron's description of the immortal (unfortunately perhaps) Mae West as "a fat old duck." . . .
>
> TV critics tend to be feature writers with fallen arches, so we are, maybe, abnormally unscientific. . . . But even measured against finer minds I think last night's Horizon was a real Lulu which might have been spoken in Urdu for all the light it shed on me. Astronomers' well intentioned attempts to simplify the issues "If we lived on a neutron star, our heads would weigh as much as a hundred large ocean liners" merely make my head loll about like (appropriately enough) Noddy.[71]

Thus, obviously, science can be funny (or can be made to seem funny) when presented for popular consumption. But perhaps this does not matter if, in being funny, the entertainment arouses the attention and interest of a mass audience. There can be little doubt that a certain entertainment value is a considerable asset, and the "fictionalization" of the implications of scientific and technological developments (as in the best kinds of science fiction) is probably the best way of alerting the mass of society to these implications. But this is too large an issue to deal with in summary fashion here, and we may content ourselves with noting that the insights offered by science fiction may at last be taken seriously—now that they can be subsumed under the "scientific" rubric of futurology!

Thus, leaving aside the question of the media's potential role in any purposive programs of mass education, it is clear that the present role of

---

69 The currently falling status of science in the community may well be matched by a reduced appeal to science in advertising—perhaps an increasing tendency to evoke sexual rather than scientific images. Such a study could provide an interesting "status meter."

70 Kreighbaum, *Mass Media*, p. 21.

71 Nancy Banks-Smith, *The Guardian*, November 23, 1971.

these organs of mass communication is a complex one, of which science-as-news forms only a part. References to science in the media transmit images as well as facts; they perpetuate stereotypes and destroy them; they reflect (and may marginally change) the status of science in society. Though the provision of factual material may determine the citizen's ability to participate knowledgeably in formulation of scientific, technological policy in the *short term,* these other less tangible communication processes may prove more important—for they may serve to determine his *desire* to participate.

INFORMATION FOR THE CITIZEN

Following Hagstrom, sociologists have tended to regard popularization of science as "illegitimate" for the scientist. Supposedly indicative of a weakened commitment to the scientific community, such behavior has been regarded as evincing negative sanctions from the community. I hope the preceding discussion has shown that such a view is oversimplistic. In the first place, we must distinguish between the pedagogic function of popularization, and what may be called its "political" function. The first describes presentation of traditional scientific theories, experiments, or hypotheses—hallowed by time and central to scientific paradigms of this or a previous age—to a nonprofessional audience. Effectively, this is teaching. No penalties attach to teaching these consecrated precepts of science, although such activity begets no professional credit either. Most academic scientists hold the view that research and (advanced) teaching are inextricably linked. The presentation of advanced research findings is rather different. All that I have had to say is compatible with the view that scientific norms require research presented for popular consumption to be first *validated* by professional consensus. Popularization implies invocation of the "authority" of science. The popularizer, speaking on behalf of science, is representing the scientific community to the general public. It is required that his statement (facts and theories) has been legitimated by prior submission to the scientific community. Moreover, the individual himself is supposed to have been approved as a fit "representative."

The scientific community is concerned about its own representation, and this concern is translated into rules, as well as institutionalized in the agencies (such as AAAS), which constitute the "diplomatic service" of science. Scientists speaking on behalf of the community must be of high repute, acquiring the right to act as spokesmen from their status. This is recognized by people in the field: hence, the preference of young scientists (in Boltanski and Maldidier's study) to talk about traditional science impersonally, even anonymously. The scientific elite adopts a much more positive attitude toward popularization and toward scientific journalists. Just as they are asked by government to sit on committees, so its members are called upon by the organs of mass communication. Elite scientists recognize that both tasks are part of their duty in representing the scientific community. Both are essentially political, deriving, in part at least, from a need to secure the well-being of the scientific enterprise.

Professional norms, norms of service, also dictate concern with presentation of scientific information to the public. The growth of what I have called "revitalized professionalism" is indicative of the growing salience of norms of this kind, springing from an increasing commitment to the idea of "science in the interests of society." This professional orientation eschews the militant or directly political attainment of objectives which may be held in common with more activist groups. The methods appropriate to the true professional are limited by the dictates of professionalism. In science, this may be translated into institutionalized concern with provision of information on scientific and technological issues to the public. Such professional commitment implies that diffusion of scientific understanding cannot be restricted to the elite "representatives" of science. All scientists with access to findings of social significance have an obligation to make them widely available.

To a great extent, however, the effectiveness of dissemination depends (in short and medium term, at least) upon the mass media—local and national, printed, oral and visual. It is principally through the media that those interested, or whose interest may be aroused, become aware of developments in science or technology which the scientific community (or anyone else) wishes to bring to their attention. And yet we have discovered that this mediatory role is complex, and cannot be performed precisely as the best-intentioned scientists may wish. Science enters the mass media most apparently as news, and the focus of journalistic interest in science is the "great breakthrough." Thus, the picture of science which the uninformed reader might obtain from his daily paper or television set would be distorted. But the fact is that the "uninformed" reader is unlikely to be at all interested in science news—except of the (literally!) most earth-shattering kind. If his attention is to be won, it must be with those eye-catching, distorting headlines which so annoy the scientist.

Whether or not the mass media are presenting a full picture of scientific progress is uncertain: the apparent doubling in editorial allocations of space may no more than keep pace with doublings in production of scientific papers. Moreover, science news is not equally *available* to all. Its availability seems to depend upon concentration of science within a given locality, and this may be as true internationally as (and we have seen this) between the states of America. Thus, Ireland has little science journalism. There is no equivalent of *Scientific American* or the *New Scientist*. "This lack of balanced science journalism is a consequence, rather than a cause, of the lack of any national scientific consciousness, expressed either in a periodical or in an event." [72] Moreover, concern with "major breakthroughs" implies that science-as-news will frequently relate to discoveries made elsewhere. Ruefully, the science correspondent of the *Irish Times* pointed out that scientific journalism tends to follow international "big" science and technology: "The space race, from the first Sputnik to the moon landing, has called forth a spate of lay-oriented articles by scientists." [23] And this in a country which spends nearly half its total research and development budget on agricultural R & D! So we must recognize that, while those in areas of high "scientific consciousness" may have available a full, balanced diet of science news, those in other places may be offered a diet both sparse and irrelevant to their needs.

In the last resort, however, popular concern must be acknowledged as of primary importance, and the media's role in stimulating concern (closely related to interest) is not bound by factual content. Entertainment, advertisement, presentation of fact or image of science may bear heavily upon attitude and interest and, ultimately, these factors will determine both the overtly scientific content of the media and the impact which such presentations may produce.

72 Dr. Roy Johnston, "Science and the Irish," *Irish Times* (Dublin), February 3, 1970.

73 *Ibid.*

chapter 8

Innovation and Society

SCIENTISTS, CITIZENS, AND DECISION MAKING

In earlier chapters, I have dealt with some ways in which scientists enter into the political process in the United States and Britain. I examined the scientist's role as advisor within government, pointing to some ambiguities inherent within it. I discussed grass roots manifestations of institutionalized political involvement: forms of scientific unionism in the United Kingdom, much with a conservative bias toward "status protection"; scientific pressure groups in the United States, some of which, like the Federation of American Scientist, work through orthodox political channels, while others (like SESPA) are concerned with awakening the conscience of the scientific community and the nation. I suggested that these grass roots movements were the result, in the first instance, of rejection of a semblance of an autonomous, uninvolved scientific role—which had, in any case, been irremediably compromised by political pressures.

Going on from there, some scientists seem to have identified with the working class, in either its radical or conservative guises, while others have gravitated toward a new (to them) conception of their own professional self-image. That is to say, some adherents of this scientific professionalism have recognized that laying a claim to professional status has traditionally entailed acceptance of a primary loyalty to society. Thus, we have a revitalized scientific professionalism, characteristic of such expanding organizations as the Society for Social Responsibility in Science, which has also given rise to the committees on "science and society" formed by a number of professional associations (e.g., the American Chemical Society's Committee on Chemistry and Public Affairs).[1] This conception has been fed by Ralph Nader and his associates, with their conception of "whistle blowing." Professionalism, radical professionalism of this kind included, affirms the primacy of the popular will, denying the right of the professional (alone or in concert) to impose his own political decisions. Inherent within it is the belief that society must be both served by and in charge of science.

This, in turn, implies that scientists must help the individual and society to understand the technological and social implications of research programs and to decide upon preferences. And this is neither more nor less than the role scientists are supposed to fill in executive decision making. (Politicians expect their scientific advisors to help *them* understand the implications of research programs, then to stand aside and allow preferences to be decided by due processes of political choice.) Such an informational role with respect to society has not only been accepted by as representative a body as the American Association for the Advancement of Science but, in addition, has given rise to a "scientists' information movement" which has close ties with AAAS but is locally organized. A similar situation may be emerging in the United Kingdom, Australia, and elsewhere. The motto of the scientists' information movement—"the scientist informs, the citizen decides"—may be taken as descriptive of this whole philosophy. In the following pages I shall analyze application of this view of the scientific role in practical politics.

In the last chapter I dealt with the scientific community's commitment to the informational role and suggested that, however great the commitment, scientists were forced to rely upon the mass media to directly reach a mass public. Research by Wilbur Schramm showed that it was from the daily press that most people learned of new developments in science and technology—if they did learn. But there have been problems in interpretation of such developments by the media in spite of the growth of a professional group of science writers, and the growth of editorial allocations of space to science news. Some scientists, conscious only of imperfections in journalistic interpretation, remain unconvinced of the value of popularization of the results and implications of their work. Second, newspapers are involved mainly with newsworthy science: journalistic science is a series of unrelated breakthroughs. Small discoveries, which may be of greater importance (particularly in the local context), are ignored, with attention focused upon the

---

1 This "revitalization" of the professionalism concept is not confined to the sciences. In the United States a radical health organization, the Medical Committee for Human Rights (MCHR), with some 20,000 members in forty local chapters, is also seeking to redefine the service role through practice.

glamorous successes of, say, space science and technology. Although straightforward misreporting is probably rare, differences in interpretation, omission of qualificatory statements, and changes in emphasis are not. These follow largely from attempting to relate what is being reported to the day-to-day experience of readers, and so to stimulate the interest of a public little concerned with purely intellectual achievement. Scientists' failure to point out the most significant implications of their work leaves the way open for at best exaggeration.

Let us assume an accuracy in reporting which may eventually be guaranteed by a still more expert corps of science journalists and a still more sympathetic community of scientists. To what extent, then, have we catered to involvement of the public in decision making in technologically based areas of public policy? Not very much. The exposure of segments of the public to science news still varies enormously. There are indications of variations on a geographic basis in the United States: science news follows scientific activity. In the United Kingdom, where major newspapers are national in circulation, there are differences on other grounds, which no doubt exist in the United States as well. Newspapers appealing to a more highly educated audience carry much more science news than those of mass appeal. So, irrespective of *individual* interest, highly educated people living in areas of high scientific activity have much greater *access* (or exposure) to scientific news than less educated people living in areas of slight scientific activity. But, in the last resort, the crucial variable is interest: no one is forced to read all that his newspaper contains: you can lead the horse to the water, but you can't make it drink.

Interest and education are crucial in determining an individual's awareness of recent scientific developments—the latter being a partial determinant of the former (as Schramm has shown). Countries, and regions within countries, differ in their concentrations of the scientifically educated, and hence in the likely interest of their populations in scientific developments. The data in Tables 8–1 and 8–2 are indicative of the comparative situations obtaining: in the first place, between countries (developed and less developed); in the second, between regions of the United States. Whichever criteria are adopted, and Table 8–1 uses four, it is apparent that only a very small proportion of the population in most countries (10 percent at best) may be expected to feel much interest in scientific issues.

There is evidence of a different sort on levels of awareness and understanding on scientific issues among the population-at-large.

A poll of the American population conducted by the Survey Research Center in 1957 provides a useful source of information.[2] Early in 1957, just

2 Survey Research Center, University of Michigan, *The Public Impact of Science in the Mass Media* (Ann Arbor, 1958).

before Sputnik, 54% of the population had never heard of space satellites, 33% of the population had never heard of radioactivity, and 26% of the population had never heard of fluoridation. Of the 67 percent claiming to "have heard" of radioactivity, 2 percent had misinformation; 11 percent had no information at all; 25 percent could make only the vaguest generalizations (e.g., "fog from the Bomb"); 21 percent had some (nontechnical) information and could talk in terms of "rays," "radium"; 7 percent had any real factual understanding of radiation, its causes, and its effects. About a year later a survey was conducted in the United Kingdom on knowledge of the concept of "evolution" among London television viewers (when two-thirds of British families owned a TV set).[3]

> Some ⅔ of the sample had some knowledge of the concept, and could define it in terms of "change," "growth": many specified change in living things. ⅓ could volunteer no information at all.
> Viewers were asked *whom* they associated with the theory of evolution (which was then defined for them). One in three could give no name, but the name most commonly given (by ⅓) was that of Darwin. "A few mentioned Huxley—but as many named Einstein." Very few had any idea whence, or how, Darwin had obtained his ideas.
> About 1/6 seemed aware that there was (or had been) some dispute between scientists and the church over evolutionary theory. ⅔ said they believed in evolution, 1/6 said they did not, the others could not say.

A more sophisticated question probed, by means of multiple choice, into what interviewees understood by evolution. The question asked can be summarized thus:

> Giraffes have long necks which enable them to feed on the leaves of trees
>
> 1. because they have been stretching them for many generations reaching for the leaves. (This would imply Lamarckian evolution.)
> 2. because the first giraffes created had long necks. (This would be called the theory of Special Creation.)
> 3. because longer necked giraffes gradually replaced shorter necked ones. (This would be Darwinian evolutionary theory.)

The responses are intriguing:

Explanation	% Agreeing	% Disagreeing	Don't Know
Lamarckian	16	50	34
Special Creation	50	14	36
Darwinian	33	14	53

3 British Broadcasting Corporation, Audience Research Department, *Evolution—A Pre-broadcast Study of the Knowledge and Attitudes of the Viewing Public* (London: mimeographed, 1958).

So the "special creation" theory won handsomely. So much for Huxley's propagandizing efforts one hundred years ago! The evolution study concludes, "If it were necessary to visualize an 'average viewer' in this connection, it would not be far from the truth to describe him thus: He 'believes

TABLE 8–1
Some Characteristics of the Work Forces of Certain Countries (ca. 1960)

Country	Definition by employment[a]		Definition by education[b]	
	Professional labor force of % of total	*Scientific labor force as % of total*	*All graduates as % of labor force*	*Science and technology graduates as % of labor force*
U.S.A.	17.3	4.7	7.6	1.8
France	11.9	5.6	2.7	0.7
Germany	10.8	5.6	3.9	1.4
Japan	7.2	3.0	4.7	0.8
Sweden	13.7	8.5	2.1	0.7
U.K.	11.3	6.0	2.8	1.0
USSR	10.8	5.9		
Israel	17.4	7.3		
Argentina	8.6	3.5		
Chile	6.8	3.4		
Egypt	4.3	1.7		
Turkey	4.4	1.0		
Pakistan	1.8	0.7		

[a] "Professional work force" is defined as major groups 0 and 1 of the ISCO code; "scientific work force" as sub-groups 0–0, 0–1, 0–2, 0–3, 0–4, 0–5, 0–X, and ½ (0–6). Data are taken from *Statistics of the Occupational and Educational Structure of the Labour Force in 53 Countries* (Paris: OECD, 1969).
[b] Data from *Gaps in Technology—Analytical Report* (Paris: OECD, 1970) Table 1, p 18.

in' Evolution which for him means not much more than that 'Man has descended from monkeys.' He recognizes that scientific knowledge in this matter is as yet incomplete and that his own is, to say the least, sketchy." True understanding, it seems, lags substantially behind awareness of scientific concepts.

Rapid changes in levels of awareness can come about and, when they do, it seems more than likely that the mass media have been partially responsible. Thus, somewhat before Sputnik, 54 percent of the United States population had never heard of space satellites; one year later this percentage was down to only 8. Reviewing an international survey of awareness of the satellite launching, the political scientist Gabriel Almond wrote, "Almost every respondent in the countries surveyed was aware of the launching of the

TABLE 8–2
Distribution of Qualified Scientists and Engineers (Q.S.E.) in American Regions, and Approximate Population-Density

Region	States Included	Number of Employed Scientists and Engineers (000)	Population (000)	Density of QSE in Population (per 000)
New England	Connecticut, Maine, Massachusetts, New Hampshire, Rhode Island, Vermont	76.7	11,412	6.7
Middle Atlantic	New Jersey, New York, Pennsylvania	235.5	36,903	6.4
East North Central	Illinois, Indiana, Michigan, Ohio, Wisconsin	223.9	39,685	5.6
West North Central	Iowa, Kansas, Minnesota, Missouri, Nebraska, N. and S. Dakota	64.6	16,043	4.0
South Atlantic	Delaware, D.C., Georgia, Maryland, N. Carolina, S. Carolina, Virginia, W. Virginia	122.4	30,077	4.1
East South Central	Alabama, Kentucky, Mississippi, Tennessee	38.4	13,113	2.9
West South Central	Arkansas, Louisiana, Oklahoma, Texas	78.6	19,234	4.1
Mountain	Arizona, Colorado, Idaho, Montana, Nevada, New Mexico, Utah, Wyoming	43.9	7,933	5.5
Pacific	Alaska, California, Hawaii, Oregon, Washington	188.8	25,560	7.4

SOURCE: Data for scientists (relating to 1962) and engineers (relating to 1960) taken from National Science Foundation, *Scientific and Technical Manpower Resources* (Washington, D.C.: N.S.F., 1964), Tables IV–2 and IV–6. Population data for U.S. states relate to 1968. The density estimates are thus only a broad approximation.

first satellite and that Russia had launched it. . . . The only other event in recent history which can match Sputnik in general public awareness was the explosion of the atom bomb in 1945." [4] But changes in understanding are much less sudden: the percentage of the U.S. population having any conception of, for example, the functions of space satellites rose in the same twelve-month period from 20 percent to only 27 percent. Krieghbaum suggests that the media can affect levels of public understanding, but only of an issue consistently in the news over a long period of time (e.g., seven to eight years).

It seems apparent that the populations of even the most highly educated and technologically advanced countries are infested by sizeable pockets of ignorance on scientific and technological matters. Levels of awareness of issues receiving substantial coverage from the media can be affected in a relatively short time, but little change results in understanding. And, outside these space spectaculars, people are not very interested: neither their education nor their interpretation of day-to-day and occupational experiences suggests any real relevance for such matters. It may be passionately argued that legislative bodies and the citizenry require more scientific advice, and that the scientific community should organize to provide such advice. The monopoly of the executive over such counsel may be decried. But interest and demand (as distinct from need) for scientific advice falls off rapidly from government to legislature to society-at-large. And so too, it may be argued, does the provision made by political systems in most countries for using this advice. Certainly, the sorts of issues with which we normally associate the need for scientific and technical advice have been the unchallenged concerns of national governments (weapons systems, space explorations); to which the "man-in-the-street" 's contribution has been marginal. But, as the focus of scientific policy changes from military and purely economic targets toward social ones, isolation of this arena of political debate from the ordinary man is reduced. And, increasingly, the scientist is called upon to demonstrate the relevance of his work for social objectives in order to obtain needed government support. The loci of power determining priorities and distribution of resources in the social field (housing, health, education, and so on) are very different from those obtaining, for example, in the military field.

Describing the American situation, political scientist Harvey M. Sapolsky has written: "Since the responsibility to act on social problems, unlike the case of military problems, is shared among levels of government in the American federal system, scientists have come to recognize that they must deal with state governments if they are to contribute to the solution of

---

4 G. Almond, "Public Opinion and the Development of Space Technology," *Public Opinion Quarterly,* 24 (1960), 573.

these problems." [5] In other words, the new emphases on scientific policy, the increasing tendency to regard science as a means to social goals rather than either purely scientific or purely military or economic goals, may be seen as having two sorts of implications. On the one hand, it has implications for the scientist's role in the decision-making process (as Sapolsky has suggested); on the other, for the potential influence over decisions exerted by the man-in-the-street. And, indeed, the need of American states for scientific advice has increased over the past years; since New York State established the first scientific advisory committee at the state level (in 1959), a majority of states have followed suit: in 1969 forty-seven had "some formally designated person or group responsible for scientific advice." [6] Witness to the states' role in performance of science-related activities is borne by their employment of technical personnel: in 1967 the states employed 20,600 scientists, 34,200 engineers, 41,000 health professionals, and 61,900 technicians.[7] Moreover, science- and technology-based programs in the social field are the responsibility of still smaller units: "Scientific and technological innovations affecting urban development must, to some degree, be carried out by the 90,000 local governments in the United States. Many of these governments do not have the financial, technical, and political capabilities to make effective use of technical innovation in carrying out their functions." [8]

My point is relatively simple. The social problems, depending in part for their solutions upon the contributions of science and technology, are not the sole responsibility of national governments. Decisions are made at many levels in the structure of political authority. Therefore, the potential (probable) involvement of the "average citizen" is likely to be greater than it would be in determination of (e.g.) military priorities to which science has in the past been relevant. At the same time not only must the need, and perhaps the demand, for scientific advice spread down the political hierarchy, but scientists needs must establish new relationships with the decision-making system in order to preserve a proportionate measure of influence.

It is not necessary to wait upon some emergent utopia in which, by instant referendum or another means discussed at the beginning of the previous chapter, the citizen will be allowed a greater say in decision making, to see how in practice this role may be acted out. As suggested in previous para-

---

5 Harvey M. Sapolsky, "Science Policy in American State Government," *Minerva*, 9, No. 3 (1971), 321–348.

6 *Ibid.*

7 *Ibid.*

8 J. D. Carroll, "Science and the City: The Question of Authority," *Science*, 163 (February 28, 1969), 902–911. (Copyright 1969 by the American Association for the Advancement of Science.)

graphs, decisions on technological innovation are made at many levels of political authority—at many places, in many institutions. A few are listed below.

TABLE 8–3
Innovations and Innovators

Type of Innovation	Innovating Group	Process of Concensus
I. New drug New kind of seed corn Rejection of non-biodegradable detergents or plastics	doctors farmers housewives	primary group inter- actions[a]
II. Fluoridation of water supplies Introduction of new educational technologies in schools	local communities	local politics
III. Elimination of colorado beetle or other pests Construction of SST Introduction of nuclear power generation	state or country	state or national politics
IV. Elimination of DDT Elimination of chemical or biological weapons Weather control	many countries	inter- national diplomacy

[a] The adoption of innovations by groups of this kind, the role of sources of technical information, and the importance of the interactive processes within primary groups of innovators have been much studied by sociologists. But the conceptual framework within which they have been studied has not been regarded as contiguous wtih that of the sociology of science. For a review of the relevant literature, see Everett Rogers, *The Diffusion of Innovations* (New York: Free Press, 1962), and the second edition published (with F. L. Shoemaker) under the title *Communication of Innovations* (1972).

Innovations range from those the individual may create for himself, which may assume considerable importance (e.g., to forego private transport systems which pollute the atmosphere, to avoid nonreturnable bottles and foodstuffs containing undesirable concentrations of mercury or cyclamates), to those involving only countries acting in international concert, upon which the individual may exert little influence. In the second part of this chapter, I select one area of "technological innovation" upon which the citizen has been allowed a considerable measure of control, in order to examine the exercise of that control and the role of scientists and related professionals within this kind of decision making. I shall show that the professionals *have* tended to act on the basis of "the scientist informing, the citizen deciding," and will then examine receipt of this information or advice by the public.

POPULAR DECISION MAKING: THE CASE OF FLUORIDATION

Controlled fluoridation is the addition of one part of fluoride to 1 million parts of water. Research over the past forty years has shown—to the satisfaction of most scientific, medical, and dental organizations—that controlled fluoridation can reduce tooth decay in children by as much as 60 percent and that it is not harmful to health. Thus, in the United States the National Academy of Sciences, the American Dental Association, and the American Medical Association have affirmed their support of fluoridation of water supplies; in the United Kingdom the British medical and dental associations have been among the organizations which have adopted similar positions. Professional opinion is not unanimous, it should be noted: many individual scientists, doctors, and dentists in both countries have taken a contrary view.

The case for fluoridation is founded largely upon public health statistics, in the first instance, upon comparison of incidence of dental decay in communities whose natural water supplies contain different quantities of fluorides. The first such study was carried out by Dean and collaborators in 1942: they examined the tooth decay in 7,257 white children aged 12 to 14 years in twenty-one cities in four American states, finding an inverse relationship between the fluoride content of the water and the incidence of dental caries. There have been many similar studies, and in later years pilot tests in a few communities. The precise mode of action of fluoride ions is uncertain, and a number of suggestions have been put forward.[9] It is, established that concentrations of fluoride in excess of one part per million (ppm) can cause severe mottling of teeth, and that at very high concentrations enzymatic processes in the body may be inhibited, bones may be affected, and there may be general fluoride poisoning.[10] So supporters have had to demonstrate that the concentration can be accurately controlled, while admitting that a small percentage (say 10 percent) may suffer adverse effects even at optimum levels of addition. Opponents of fluoridation—

---

9 "Fluoride may act in two ways, which are not mutually exclusive.
1. By rendering the enamel less soluble in acid.
2. By the known anti-bacterial and anti-enzyme properties which fluorides possess. The effect of acid on the surface of the enamel could be to release fluoride ions, which, especially if confined by plaque formation, could reach a high concentration. The process being reversible, the fluoride ions could later, when the pH rose sufficiently, be re-incorporated in the enamel." British Dental Association, *Fluoridation of Water Supplies* (London: May 1969).

10 *Ibid.*

among whom scientists, doctors, and dentists are but a minority—have questioned interpretation of the all-important statistics, alleging defects in statisical controls which establish the independent action of the fluorides. They have suggested that any apparent improvement is simply delay in the incidence of decay and, moreover, that fluoridation of water supplies is "compulsory mass medication," and therefore a dangerous precedent; that it is a form of pollution (since fluoride emitted by aluminum smelters is widely so regarded); that this compulsory medication is a result of political pressure, generated by commercial interests seeking new, profitable outlets for this material. Antifluoridationists argue that bulk addition to the water supply is neither the cheapest nor the best means of administering the material. The argument on the grounds of principle has been perhaps most authoritatively stated by Harold P. Green of the National Law Center of George Washington University, who has written:

> Why should anyone be distressed that the public, for whatever reason, rejects technological benefits? . . . Why the great hurry? . . . dental caries is a personal disease which does not spread . . . (and for which there are) alternatives (remedies) . . . without forcing medication upon those who don't want it. There is a question of principle involved which transcends the fluoridation controversy. . . . Will we, a decade hence, be compelled to accept additives to the public water supplies to reduce fertility, sharpen mental processes or tranquilize the population, because the legislature so decrees.[11]

The political controversy which surrounded—and to some extent still surrounds—fluoridation has not been founded upon objective comparison of the two sets of arguments. Interest in the drama depends not solely upon the script. The standing and credibility of the actors; the way they speak their lines; the arena in which the performance takes place—all are relevant. And, needless to say, the impact of a line depends upon the character from whose mouth it issues. The roles of scientists and health professionals in the fluoridation controversy are not to be described solely in terms of research findings and statistical analyses reported in the professional literature. As I have said, the health professional associations almost unanimously spoke in favor of fluoridation: the scientific world has been vocal in its advocacy of fluoridation, and there has been little organized support for the opposite position among people in the world. At the local level, at which decisions have been made on this issue in both the United States and Britain, professional associations have spoken in favor, frequently initiating debate leading

---

11 Harold P. Green, Letters, *Science,* 163 (1969), 17. (Copyright 1969 by the American Association for the Advancement of Science.)

to adoption or rejection. Studies of local situations have thrown more light upon roles of these professional fluoridationists. "The pro-fluoridation leaders, mainly dentists, felt that the weight of scientific evidence and authority was so strong that they did not need to campaign intensively." [12] Having studied a community in which fluoride was rejected, these authors go on to criticize scientists and health professionals for having "relied too heavily on the fiat of organized science" and for unwisely dismissing their opponents as crackpots. Subsequently, an attempt was made to explain the behavior of professionals. Raulet suggested that the involvement of this group in the local decision-making process resulted in role strain.[13] "Where the campaign to fluoridate the local water supply is carried by a partisan group made up mainly of health professionals, it seems likely that the professional expert role and the partisan role will conflict sharply." [14]

Proponents of fluoridation presented their case primarily in the form of a public health program. Since health education is part of the traditional professional role of this group of people, it appeared to them an appropriate way to deal with what, in scientific terms, may have been a run-of-the-mill issue. In consequence, efforts were made not to engage in political debate, since to make fluoridation a political issue would mean admitting that expert professional opinion was no more privileged than lay opinion. Worse still, if antifluoridation doctors or dentists were around, since political debate under these circumstances would require a public admission that there was a scientifically respectable opposition. Thus, according to this analysis, the debate preceding a referendum in an American town placed doctors and dentists in a difficult position, since the behavior they regarded as appropriate to their professional roles contrasted sharply with that expected in referendum contests. Since they avoided political debate, they were accused by antifluoridationists of acting in accordance with material self-interest, of collusion with the chemical industry, and, by supposedly trying to suppress antifluoridationist arguments, of violating norms of democratic behavior. Thus, they found their intentions impugned and their claims to authoritativeness unacknowledged. In fact, details of the roles played by health professionals differ between the United States and Britain, since in the former the referendum has been the rule, in the latter it has not, and decisions have been made largely by local councils.

Raulet's study of two Massachusetts towns, to which I have referred,

12 Bernard Mausner and Judith Mausner, "A Study of the Anti-Scientific Attitude," *Scientific American,* 192 (1955), 35–39.

13 H. M. Raulet, "The Health Profession and the Fluoridation Issue," *Journal of Social Issues,* 17 (1961).

14 *Ibid.*

may be compared with one of the decision process in Hull, England, carried out by Brier [15]. In Hull the initial proposal that the water supply be fluoridated came from the town's Medical Officer of Health, supported by the local branch of the British Medical Association (BMA), which, however, reserved its position upon technical feasibility, financial and administrative matters. Throughout, the Medical Officer of Health (MOH) was the main supporter of fluoridation but, even so, his role was not an activist one, since he acted principally as expert advisor of the lay councillors. Moreover, and upon the advice of the MOH, the local branches of BMA and BDA (British Dental Association) took little interest in council deliberations beyond formal expression of their views, and in no sense did they act as pressure groups. (Brier suggests that this may have been the result of their reluctance to jeopardize their bargaining position on other matters with the council.) It is not clear how typical was the Hull situation, and Brier suggests that other MOH's have tried to act *both* as expert advisor and protagonist, striving to lead local professional bodies into more activist positions. It seems possible that the role strain which Raulet observed in health professionals in his two Massachusetts cities was in part a result of the decision-making-by-referendum system. If decisions are made in committee, and if established means exist for feeding expert advice into the committee's deliberations, then the expert role and the protagonist role may be relatively compatible. (After all, expert advisors are frequently pushing a particular viewpoint or interest, as we saw in Chapter 6.)

But if the debate is conducted in public, if the final decision rests with the public-at-large, the situation may be rather different. In the first place, as Raulet found, though the professionals were anxious to demonstrate their support, they lacked institutionalized channels for so doing, and leadership tended to be emergent. Second, the attempt to persuade to a specific viewpoint must be in far more strident tones, far more emphatic, if the slightest impression is to register upon an often apathetic public. Compatibility with professional objectivity, on grounds of style as well as content, becomes more difficult. And, finally, though this may be presumed more with doctors and dentists than with scientists, professionals may be more conscious of their purely professional obligations in dealing with the public than in dealing with a political committee. Thus, the professional objectivity which they have been taught must characterize their dealings with the public (potential clients) may be a substantially more formidable barrier to any exhortatory position in dealing directly with the public (decision makers) than in dealing with its elected representatives. On all these grounds, it seems

15   A. P. Brier, "The Decision Making Process in Local Government: A Case Study of Fluoridation in Hull," *Public Administration* (Summer 1970).

to me, the position of the professional in popular decision processes may be more difficult and conflict-ridden than the position defined for him in executive decision making.

Now let us focus our attention upon the referendum process by which fluoridation has been accepted or rejected in many American communities. Between November 1950 and December 1966 there were 952 referenda on the fluoridation issue in the United States, and in 566 fluoridation was rejected.[16] The fact is, as Sapolsky has pointed out, that where fluoridation has been adopted by an American community it has usually been via a different kind of decision process.[17] Table 8–4 shows an extremely good correlation between rate of adoption by communities of a given size and size of communities: 60 percent of communities of 1,000,000+ considering the issue have adopted fluoridation—compared to only 14 percent of those between 1,000 and 2,500. Table 8–5 shows some voting figures for two referenda held in 1952.

It seems possible that the correlation demonstrated by Table 8–4 may be explained in terms of the mode of government typifying each group of communities. Crain and Rosenthal reviewed the variables seeming to determine the mode of decision making typically adopted in considering fluoridation.[18] They concluded that government structures permitting public participation (more referenda) tended to adopt fluoridation rather rarely. On the other hand, those which restricted public participation and involved few referenda tended to adopt fluoridation quite frequently—by administrative decision. That is, towns with strong executives, formally less democratic, were far more likely to adopt fluoridation, as advised by the public health authorities. Sapolsky suggests that the larger cities, frequently tending to opt for fluoridating their water supplies, may most often make use of a government structure limiting direct public participation. The strong executive, particularly in towns with a strong mayor, rarely consults the voters. To summarize, almost all decisions to fluoridate water have been made by administrative order; most referenda have rejected the process.

Does this imply profound contradiction between technological progress (if fluoridation is regarded as such) and democracy (if referendum decision making is regarded as essential to pure democracy)? This is Sapolsky's view. Making the point we have made earlier, he suggests that shifts in scientific priorities toward social problem–oriented research are likely to lead to an

16 H. Sapolsky, "Science, Voters, and the Fluoridation Controversy," *Science,* vol. 162 (1968).

17 *Ibid.*

18 R. Crain, and D. Rosenthal, "Structure and Value in Local Political Systems: the Case of Fluoridation Decisions," *Journal of Politics,* 28 (1966), 169–196.

TABLE 8–4

Authorization of Fluoridation

Population Range of Community	Number of Communities	Using Controlled Fluoridation	As Percentage of All Communities of Same Size	Source of Authorization to Fluoridate			
				Govt. Body Alone	Refer- endum	Utilities Com- mission	Other
1,000,000+	5	3	60.0	3	—	—	—
500,000 to 999,999	16	8	50.0	7	1	—	—
250,000 to 499,999	30	13	43.3	12	—	—	4
100,000 to 249,999	81	35	43.3	29	4	4	1
50,000 to 99,999	201	82	40.8	67	8	4	3
25,000 to 49,999	432	167	38.7	129	13	4	21
10,000 to 24,999	1,134	465	41.0	363	43	16	43
5,000 to 9,999	1,394	505	36.2	378	40	21	66
2,500 to 4,999	2,152	507	23.6	389	42	16	60
1,000 to 2,499	4,471	620	13.9	490	39	27	64
Under 1,000 and unspecified	10,677	740	6.9	568	26	42	104
	20,593	3,145	15.3	2,435	216	131	363

SOURCE: H. Sapolsky, "Science, Voters, and the Fluoridation Controversy." *Science*, 163, (October 25, 1968), 427–433. (Copyright 1968 by the American Association for the Advancement of Science.)

TABLE 8-5
Voting in Two Referenda

	Seattle	San Francisco
Voting for	44,814	114,125
Voting against	86,230	88,377
Population (1950)	467,591	755,357

increasing number of comparable issues requiring local government approval, which he regards as subject to frustration by referendum.[19] Some writers, like the Mausners, have "blamed" scientists and professionals for failing to recognize the appeal of antifluoridation literature.[20] (In Hull, Brier found that antifluoridation literature put out by the National Pure Water Association was the most common source of information on the subject: it became accepted as an authoritative source.) Sapolsky "blames" public ignorance. The public must be educated to understand the issues, a view from which it is hard to dissent, since issues such as this are too complex for them to resolve. They need education to utilize the advice of scientific experts. But is this education for understanding or education for acceptance of the greater capabilities of the professionals: education for submission? Under present circumstances, Sapolsky prefers decisions to be made by administrative bodies with their "greater capacity to distinguish among the experts." This is not an altogether felicitous view. But the fact remains that the general public has not shown any great enthusiasm for fluoridation: whether this has wider implications for its attitudes toward new technologies remains to be seen. But the possible reasons behind this lack of enthusiasm are worthy of examination.

Considerable interest has attached to explaining the usual rejection of fluoridation in referendum contests. The Mausners interviewed a sample of potential voters in Northampton, Massachusetts, attempting to ascertain the disribution of antifluoridation attitudes, as well as to determine the relevance of various antifluoridation arguments to opposition.[21] They concluded that resistance was to be found mainly among the older, lower income bracket and among less well-educated members of the community. Many opponents, the Mausners found, felt that public health officials and dentists, together with the chemical industry, were conspiring to impose the measure on the public. Their general explanation is in terms of this fear of conspiracy, this "tendency to perceive the world as menacing." The authors were struck

19   Sapolsky, "Fluoridation Controversy."

20   Bernard Mausner and Judith Mausner, "Anti-Scientific Attitude."

21   *Ibid.*

"by the pervasive attitude of suspicion among those who opposed fluoridation. They were suspicious not only of scientific organizations but of the scientists themselves. To them, as to all those who fear the 'egghead,' it seems perfectly reasonable to suppose both that scientists would lend themselves to a conspiracy with enemies of our country, and that, at the same time, they would permit themselves to be used by a giant monopoly." They believed that fluoridation had been promoted largely for profit—the profit of dentists, the profit of the chemical industry. Scientists were regarded as part of this general conspiracy and, apparently, scientific authority was to be rejected on just those grounds. Other authors have taken a rather similar view: the antifluoridation "crusade was a political protest more against the scientific orgnizations and a government which heeded their wishes." [22]

Although there has been disagreement over the proper distinction between the views of the active opponents of fluoridation and the more widely held views of those who merely voted against fluoridation, most authors have adopted a *somewhat* similar view as to the source of opposition to fluoridation. Davis, for example, argued that the various forms in which opposition to fluoridation is expressed (fear of large and powerful organizations—such as the U.S. Public Health Service; strangeness of an unknown, untrusted drug; identification of mass medication with "socialism") may be regarded as part of a general "opinion syndrome." [23] This syndrome he discusses in terms of a preference for what is natural: a "naturalist syndrome." Scientists are the principal creators of what is unnatural and, therefore, the last people to trust. Scientists, for Davis' "naturalist," are manipulators of incomprehensible symbols, vivisectors of helpless animals, creators of unimaginably terrible weapons. Other authors have sought to equate opposition to fluoridation with concern for preservation of human rights (and this view has certainly commended itself to many intellectually vocal opponents). Others have rejected this view: "The general conclusion seems inevitable, that overt differences in ideology concerning individual rights and government intervention have little to do with the average voter's position on fluoridation (in this Cambridge precinct)." [24] (This conclusion was based upon a series of interviews conducted in a predominantly working-class area of Cambridge, Massachusetts.)

Another study (also of Cambridge) suggested that attitudes toward

22   D. R. McNeil, *The Fight for Fluoridation* (New York: Oxford University Press, 1957).

23   Morris Davis, "Community Attitudes towards Fluoridation," *Public Opinion Quarterly*, 23 (1959), 474.

24   William A. Gamson, "The Fluoridation Dialogue: Is It an Ideological Conflict?" *Public Opinion Quarterly*, 25 (1961), 527–537.

fluoridation are connected with general political views.[25] In this study of a 1953 referendum, it was shown that areas voting for fluoridation also tended to support the "more liberal" view on other issues. But on this specific relationship—equation of liberalism with approval of fluoridation—there is conflicting evidence and, in the first of the two Cambridge studies referred to above, the authors found that "small differences that do exist show the opponents of fluoridation taking more liberal positions. In the 1956 presidential election, 68 percent of the proponents voted for Eisenhower against 58 percent of the opponents." [26]

Somewhat more plausible than the simple relationship between attitude toward fluoridation and party-political view, and in a sense returning to the "conspiracy" idea introduced earlier, is the suggestion that attitudes on this issue can be explained in terms of "alienation." The idea of such a source of relevant attitudes has boasted a number of proponents, opposition to fluoridation being viewed as in some way a function of a feeling of helplessness, of inability to control the events that largely shape their lives.[27,28] It is perhaps useful to summarize the conclusions of one such study: [29] "The hypothesis that opponents will have greater feelings of helplessness and a lower sense of political efficacy than proponents is fairly well supported by the data given here. It is as if fluoridation somehow symbolized the buffeting one takes in a society where not even the water one drinks is sacrosanct. Furthermore, the leading proponents are generally professionals of high status but relatively little power, making them a particularly inviting target" —and this is especially the case when professional proponents assert the purely technical nature of the whole issue, dismissing opponents' views as mere crankiness. It must be admitted that this alienation hypothesis, although rather satisfying as an explanation, is based upon tenuous evidence.

And, as an explanation of general voting behavior (as distinct from an explanation of leading opponents' views of fluoridation), it has had its detractors. Sapolsky has argued that available interview data show no evidence for general rejection of science and scientists and, indeed, that, due to the influence of a few dissenting *scientific* organizations (e.g., the American Academy of Medicine, the National Health Foundation, the Association of American Physicians and Surgeons), many individuals have rejected

25 Thomas F. A. Plaut, "Analysis of Voting Behaviour on a Fluoridation Referendum," *Public Opinion Quarterly,* 23 (1959), 213–222.

26 Gamson, "Fluoridation Dialogue."

27 James S. Coleman, *Community Conflict* (Glencoe, Ill.: Free Press, 1957).

28 Arnold Simmel, "Relative Deprivation as an Hypothesis in Studying Community Acceptance of Fluoridation," mimeo, quoted Gamson, "Fluoridation Dialogue."

29 Gamson, "Fluoridation Dialogue."

fluoridation.[30] Also, he argues, there is no external support for the view that a substantial proportion of the community rejects science. Fluoridation, however is different from science-as-a-belief-system: it is *embodied* science, acceptance of which requires not only the acceptance of scientific authority but also of general attitudes with regard to authority and appropriate government action. I have referred to Sapolsky's conclusion that the fluoridation issue shows the imperatives of technological progress placing a limit upon the extent of allowable citizen participation. In his view, the intricacies of the choices are too great for the average voter—too formidable, therefore, for submission to popular referendum. It is not to take issue with him on this point that I refer to Sapolsky's paper at this juncture, but I believe he fails to present an alternative explanation to that offered by "alienationists." Moreover, identification of science with what the alienationists' claim is rejected seems far from obvious. Gamson considers that his statistical data show "distrust of the agents who do control (events)," but the agents to whom his questions refer are "public officials."

Research has been involved basically with the attempt to find an explanation for *rejection* of fluoridation—the *rejection* of scientific authority. In the 1950s and early 1960s, when most of this work was done, it was taken for granted by social scientists that the norm was the acceptability of scientists' dictates: the authority of science could not rationally be criticized. This is not the case today, when it is intellectually acceptable to reject science, and certainly acceptable to reject science's claims to ultimate authority. So the alienation hypothesis must be seen as a product of its time. Is it evidence that much of the U.S. population rejected science when science was looked on with unprecedented favor by politicians, presumably acting on behalf of the electorate? What does the controversy over fluoridation tell us about the proper role of scientists and professionals in popular decision making on issues involving science and technology? What does it tell us of the general public's involvement in an issue presented, in the United States at least, for popular choice?

Let us take the last question first. We have seen that the general levels of both interest in, and knowledge about, scientific matters are exceedingly low. Also, while changes in awareness of specific issues are readily brought about (so that the percentage who "have heard" of fluoridation rises rapidly), they are rarely accompanied by changes in levels of understanding, so the influence of substantial publicization of a complex scientific or technical issue will likely be limited to the superficial level. This is the background against which to judge popular responses to fluoridation referenda. What were these responses? In the first place, relatively few people bothered

---

30 Sapolsky, "Fluoridation Controversy."

to vote (in Seattle and San Francisco, the figures were around 27 percent of the population; see Table 8–5). In the second place, most referenda rejected the viewpoint which scientific findings and most scientific and medical organizations seemed to favor. Since few people are likely to have attempted (or been able) to weigh the complex statistical, scientific, and medicodental data, it may be presumed that acceptance of the fluoridationist position did not imply an inductive argument from these data in line with that of the professionals, nor did rejection imply an alternative argument from the data. Acceptance, then, must have signified acquiescence to the interpretation placed upon the data by scientists and health professionals, and to policy implications drawn from them.

Therefore, it is perhaps not too gross an oversimplification to regard popular acceptance of fluoridation as a result of either popular readiness to accept the expert, disinterested *advice* of professionals or to accept the *authority* of the scientific viewpoint, almost unconsciously. (Therefore, it is not self-evident that to approve fluoridation is to act rationally, as most investigators have assumed.) The rejection of fluoridation has found many possible explanations: fear of conspiracy; alienation from the political system; rejection of science; concern with preservation of human rights—and doubtless many more. All explanations have received some support from distinctly ambiguous data. They may be summarized:

Attitude Toward Fluoridation	May Imply
pro	1. rational acceptance of expert advice
	2. unthinking acceptance of scientific authority
anti	3. rejection of scientific advice because of supposed conflict with individual freedom of choice
	4. tacit rejection of authority of science
	5. alienation from and rejection of the political system
	6. total alienation from society: fear of conspiracy

Numbers 3 to 6 imply rejection of scientific advice: 4, 5, and 6 seek, in addition, to explain this rejection as pathological. No matter what the "ultimate explanation" (or distribution of reasons within the population), we are quite safe in interpreting popular rejection of fluoridation as rejection of (at least) the advice of most scientists and health professionals.

Consider, finally, the behavior of scientists and health professionals in this decision-making process. It is apparent that their approach to the whole issue was preeminently "professional"—or, put another way, "informational" or "advisory." The result, as we have seen, was that their advice tended to be rejected. Why, then, did they behave in this way? why was there such a pronounced reluctance to engage in the kind of political procedures which have normally categorized either referendum issues in the United States or the more politically live issues in British local politics. Various explanations have been put forward in attempts to understand this self-imposed limitation upon the political effectiveness of the professional group. It has been suggested, for example, that an attempt was made to play down the issue politically to conceal the existence of a professionally respectable opposition, or because including it as a political issue would be admitting the possibility of dissent from scientifically authoritative proof. Both interpretations may be considered aspects of the preservation of professional status. To have agreed to submit to political debate an issue seen as falling purely within professional competence would have involved compromising that status. And, while the professional is regarded as "disinterested," the political participant's motives may appropriately be questioned. Another perspective suggests conflict between the role of "disinterested professional" and that of "protagonist"; that is, between behavior normally expected of professionals, and that required of them by conditions of a referendum. The professionals, viewing the role in which they were cast by the referendum contest as in conflict with their usual professional role, rejected it. They sought to act out a *purely* professional interpretation of their role in society, rather than to adopt a more political, persuasive approach.

It is not easy to summarize conclusions of general importance deducible from the fluoridation issue alone, and previous explanations have perhaps strayed too far beyond the evidence. Moreover, there is little complementary evidence from other comparable tests of public feeling, so generalizations must be tenuous in the extreme. Nevertheless, it does appear that scientists, technologists, and related professionals concerned with ensuring democratic control of technological change and with protecting society from ill effects deriving from such change, must recognize that these two eminently fine objectives may require them to act in very different ways. The "professional" approach—to inform and to stand back—may serve democracy, but only to the extent that it overcomes society's enormous apathy toward technological issues. Yet the chances of professional advice being rejected are so high that behavior of this kind must be seen as leaving the field open for those less scrupulous, prepared to adopt a more active line: whether the issue is health, conservation, armament, or construction of an SST. Only by adopting a more activist interpretation of their proper place in the political system can scientists and technologists save society from technological disaster.

THE POLITICAL SOCIOLOGY OF SCIENCE

The problems raised in the last section delineate a convenient starting point for attempting to summarize the general argument of this book. The fluoridation issue seems to support the following generalizations. First, the scientists and health professionals (principally the latter; the reader must allow me a little license here) were anxious to act in a purely disinterested way, even though the referendum situation created a potentially activist, unusual role for them. Secondly, the general public tended to reject the largely profluoridationist advice of professionals (leaving open for the moment the question of what else they may or may not have rejected).

The first generalization is best understood in terms of commitment of groups (including scientists) to a professional conception of their role in society. (There is other evidence, outlined in earlier chapters, that scientists prefer to be regarded as professionals, rather than *sui generis*.) Their behavior under the referendum conditions may then be attributed to attempts at preserving their professional status, by concealing internal dissensus and denying the questionability of an issue which they regard as professional. Alternatively, the professionals' behavior may be seen as manifestation of the political stance proper for the professional vis-à-vis society: to advise, to inform, but not to lead. From this perspective, the contrast between the disinterested behavior in the referendum and the frequently very different behavior in closed policy debate may be understood in terms of the professional norms determining the relations of the scientist or physician with members of the public. By contrast, he is not so powerfully socialized into any specific view of his relations with bureaucrats and officials, and is freer to act as he feels the specific situation demands. It is doubtful if scientists (as distinct, perhaps, from health personnel) have usually taken their relations with the public as so clearly defined by norms setting forth their responsibilities: the scientific view of professionalism is discussed again below.

The second generalization concerns rejection of medicoscientific advice by the general public. I suggested that merely to frame this interrogatively as social scientists have done—"Why was the advice rejected?"—is to betray an important aspect of the investigator's beliefs. By framing the question in this way, social scientists demonstrate their commitment to a specific view of the social relations of science. Why should the advice of scientists and dentists be accepted, in a *political* situation, in preference to the advice of any other group with or without a vested interest? To attempt to explain rejection of this counsel as a rejection of science, or even rationality, is curious in the extreme: who expects government policy makers to accept unquestioningly the views of military, economic, or scientific advisors? Politicians are expected to attach equal or greater significance to more general considerations, or to attempt striking a balance between the frequently conflicting views of various groups of advisors. The explanation of the social scientists' preferred explanation is that, whether consciously or unconsciously, there is belief in the high status of science in industrial society. Only in light of this presumed status—to which politicians are assumed less blindly responsive—does the rejection of scientific advice become questionable. Scientific rationality, until recently, has been a most pervasive ideology (although not that alone). There is other evidence for the ideological nature of science, provided by appeal to science as means of persuasion: whether in advertising or by political groupings such as Scientists and Engineers for Johnson-Humphrey.[31] Further evidence is found in the radical critiques of Marcuse and Habermas discussed in Chapter 2. The association of science with technical progress, and the political standing of this notion of technical progress (until recently),[32] confers a distinctly political flavor upon the scientific ideology.

Let us return to the concept of scientific professionalism. The scientific role in society evolved out of growing realization by an upwardly mobile segment of the community that its own needs coincided with those of an embryonic group of "scientists." But the role was transformed by a recognition that science could be of some direct utility. Chemistry was the first branch of physical science for which such a directly economic function emerged. Though it was rarely to carry out research that nineteenth-century industrialists began to employ chemistry graduates in increasing numbers, they appreciated both that their industry (the chemical industry) was founded upon chemical research and that skilled chemists had much to contribute to its day-to-day operation.[33] Chemists became "professionalized" to attain and then to preserve the social and economic status to which they felt entitled, and to ensure reservation of the title "chemist" for those proven worthy of it. The interrelationships of these requirements were recognized long ago. Gradually, other scientific disciplines followed the same road, and the process of their successive professionalization may still be observed today.

I have argued that the effects of professionalization upon scientific societies were similar to the effects of intellectual differentiation, and evinced similar responses. Scientific institutions respond to changing economic circumstances. But the professionalism of the scientific specialties was always recognized as significantly different from the professionalism of the older fields: chemists and physicists rarely had a "client" in the sense that lawyers and physicians did and do. They were employed, principally by industrial

---

31   See chap. 2.

32   Norman J. Vig, *Science and Technology in British Politics* (Oxford: Pergamon Press, 1968), chap. 3.

33   See chap. 4.

firms, and it followed that their obligations were of a somewhat different nature: "The predominating characteristic expected of the chemist is loyalty. . . . He must of course carefully avoid talking of his work and the affairs of his employers to those who are outside the business." [34] Traditionally, the emphasis has been upon the scientist's loyalty to his employer. But the mounting awareness of the *social* implications of science and technology, particularly in postwar years, has stimulated recognition of the other side of the professional coin. Scientists have begun to appreciate that to lay claim to the status of professionalism—which inheres in the relations between the group and society as a whole—must imply acceptance by the professional of his obligations to society as a whole. Although recently Ralph Nader has publicized these obligations with his concept of "whistle-blowing," [35] their prior acceptance by important segments of the scientific community is clear. This is apparent in the postwar statements of purpose by AAAS and in the existence of the scientists' information movement.[36] Professionalism ought to imply the primacy of dedication to the well-being of society, to which the professional must subordinate both his immediate employer's demands (in spite of legal constraints) and the potential power of his own professional group. I have argued that emergent service norms may be the major source of strain for scientists in organizations (in contrast to earlier views of strain as deriving largely from purely scientific norms, norms of autonomy). Professional service norms will require a much wider commitment to diffusion of scientific knowledge than previously. Popularization can no longer be seen as the prerogative of those legitimated as suitable "representatives" of the scientific community, concerned to act politically in the interests of science.

Socioeconomic or political pressures may lead scientists, no less than other professionals, to retreat from this position. Status claims may be modified, acknowledgments of responsibility conveniently forgotten. The employed scientist may come to feel that he has more in common with nonprofessional employees than with self-employed professionals. However gradually, however reluctantly, he may turn from the genteel self-restraint of professionalism to the boisterous militancy of trade unionism. Whether he will adopt only the instrumentalism of trade union bargaining procedures, or an ideological commitment to the principles of social democratic or labor politics, or both, seems to depend upon the exact *social situation* with which he is confronted.[37]

34   R. B. Pilcher, *The Profession of Chemistry*, 2nd ed. (London: Royal Institute of Chemistry, 1927), p. 83.

35   See chap. 4.

36   See chap. 5.

37   See chap. 5.

The sociology of science must also take note of these "unionization" and "politicization" movements, and the institutions through which they work. They are a perfectly natural response to the economic and political situation in which the modern scientific enterprise must exist. We may not assume that socioeconomic change ceased to be relevant for science when the scientific role had first become established in the seventeenth century. Professionalization was a second major response of the institutions of science to economic change. The movements described in Chapter 5 are a third. If emergence of the Royal Society was a legitimate consequence of a social situation increasingly favorable to science, founding of the Federation of American Scientists is, analogously, an equally legitimate result of a less favorable situation. At the same time, we must recognize the different kinds of responses to crisis within the scientific community: "professional" and "traditional" unionism, both emphasizing the status and salary requirements of scientists; "politicization," emphasizing either inappropriateness of government policy or general perversion of the scientific enterprise.

The growing application of science, the growing belief in its potential and actual utility, has not followed solely from the increasing sophistication of private industry. It was just such a view, reinforced by an urgent need to mobilize all national resources for conduct of the war, which led the British government to create the Department of Scientific and Industrial Research in 1915 and the American government to create the National Research Council in 1916. Both first and second world wars, in their turn, served to emphasize the need of science and government for each other. Governments required not only performance of research but also advice on formulation of suitable research programs. Scientists previously unwilling or unable to try to interpret each group's needs in the councils of the other found themselves filling the new advisory roles brilliantly. In the United States, specifically, the end of hostilities did not bring about dismantling of the new advisory apparatus. A belief in the long- and short-term benefits of science, surpassing even the immediate exigencies of the cold war, produced a growth in support for the discipline extending to fields with no apparent practical applications. Programs multiplied and, to advise on allocation of very substantial sums of money, policy makers found themselves in continuing need of authoritative scientific advice. "Statesmen of science," who often seemed to spend much more time in the "corridors of power" than they ever did in the laboratory, became a recognizable species. In an earlier chapter, I dealt at length with the problems faced by these scientific advisors, and with the sources of power and influence which they may tap in dealings with their political masters.[38] Regarded usually as representatives of the scientific community, their principal weapon will often be the status, the incontrovertibility of science—

---

38   See chap. 6.

although personal relationships may assume greater importance. So fluctuations in the status of the field will seriously affect their political influence. In this basic respect, the scientific advisor differs from the career-administrator within the bureaucracy: his influence depends upon the mutual regard in which political leaders and scientific community hold one another.

So far in this summary I have suggested, implicitly, that the external relations of science have brought into being certain kinds of political roles: roles in the political drama reserved, as it were, for scientists. One derives from the ideological nature of science, allowing the individual a degree of general political influence. A second derives from the professionalism of science, requiring the scientist to help society determine how it may best harness science and technology for its welfare. A third derives from the relationship of science to executive government, and from the needs of government. Finally, I should like to outline implications of these political relationships for internal operation of the social system of science.

Politicians want their advice to be authoritative. Therefore, advisors must possess substantial experience in scientific activities, constituting expertise. (The analogy with mass circulation publications also requiring "authoritative" writing about science may be drawn here.) The advice must also be acceptable, both to the scientific community and to the politicians. The first follows from the desire for smooth external relations; the second from the desire to preserve consensus, so much a part of the administrative machine. Such requirements imply further characteristics desirable for scientific advisors. They should command wide respect within the scientific community as a whole, each in his own discipline of science. They must be prepared to work "with" their political masters, and should share their political beliefs at least to the extent that continuing dialogue is possible.

Three characteristics of incumbents for this advisory role seem to follow: scientific eminence (as indicated by membership in the National Academy of Sciences); representativeness (implying a balance between the disciplines of science); and appropriateness of political beliefs (which must lie within some broad consensual range). Policy makers may be expected to utilize these criteria implicitly in choosing science advisors.

We must recognize, however, that from within the scientific community these advisory roles assume a rather different importance. To the scientist, occupancy of such a role represents not only the political opportunity for pressing the case of his own discipline, but an important symbol, bestowing recognition of scientific achievement on him. With very few exceptions, scientists regard an invitation to serve on these committees and panels as a recognition of their contributions to science, comparable (though not on a par) with election to NAS or the Royal Society. Such recognition is central to operation of the social system of science. Scientific norms prescribe for its allocation to an individual solely in proportion to his or her contribution to the advancement of science.[39] Adoption of political criteria in selecting advisors (although but one example of deviation from norms in allocation of rewards), is therefore resented within the scientific community. Moreover, the demand for "scientific eminence" leads to selection of advisors solely from within an elite of the scientific community, thus drawing attention to, and strengthening, a division proscribed by the norms of science.

Operation of the reward system of science serves to create an elite. In part, this follows from preferential rewarding of the previously successful (Matthew Effect) and, in part, from the fact that once professional standing has been acquired it is never lost but becomes "ascribed" independently of the individual's further efforts. Some argue that this stratification is functional for science since, because the elite scientists (for example, Nobel Prize winners) are likely to choose the more exciting, innovative, fertile fields for study, the special attention which their work receives will keep science from bogging down in trivial problems. But this view seems to depend upon three basic assumptions.[40] The elite must consist of Nobelists and others of similar caliber, promoted by virtue of their ability only; *and* they must remain innovative and capable of exercising intellectual leadership. Moreover, the functionalism of stratification cannot be assessed in terms of intellectual (and voluntaristic) leadership only: note must be taken of coercive powers involved. To the elite, many specific powers of evaluation are delegated by the scientific community: the elite acts as "performance judges" in electing to honorific academies, awarding prizes, refereeing papers, and so on. Also, additional power is delegated to the elite by government: especially, power over formulation of policies and disbursement of funds. Therefore, we must make certain that, in discharging these duties, the elite is not swayed by those views and background experiences differentiating it from the scientific community-at-large.

The first assumption has been discussed in some detail. We can feel fairly sure that access to the elite is not on grounds of scientific achievement alone: factors such as academic background, institutional affiliation, and sex are highly relevant. There are grounds for doubt. Political considerations are also among these "nonuniversalistic" factors governing access to the elite. Reflecting politicians' preferences, they are especially relevant to choosing members of official scientific advisory committees. Government requirements abet stratification of the scientific system. But they do more. They *politicize* relations between the elite and the scientific community, by virtue of the former's association with a government for which many scientists may feel little sympathy. This politicization follows from the close ties binding many national academies of science to government, and from the fact that the elite

39   See chap. 2.
40   See chap. 3.

is a clear, single source of advice. When government policies are repugnant to the majority of scientists, this politicization may take on the appearance of constituting the most important dimension of stratification within science. That scientific advisors may have been uninvolved in formulation of these policies, or that they may have (secretly) sought to oppose them, may seem irrelevant in terms of perceptions of the scientific community. Political considerations may become its major source of division.

From this follows the question of representativeness. One function of the advisory machinery is to represent the scientific community in counsels of government (which may occasion substantial dissaffection). To be legitimate and a source of both political and communal satisfaction, representation must be based upon an appreciation of those divisions within the community which *members* of the community regard as of prime importance. Only thus can justice be both done and seen to be done. Traditionally, disciplinary divisions have been regarded as paramount, and "representation" on scientific committees has been held to require participation of disciplinary interests to the extent that numbers permit. Perhaps this should no longer be the case, and representation should reflect institutional, geographic, or age differences, or variations in attitude toward the social responsibility of scientists.

In conclusion, the major theme of this book has been that the social structure of modern science is highly dependent upon the social, economic, and political organization of society, and extremely sensitive to changes in this environment. It seems to follow that the very condition of being a scientist places the individual within a network of relationships to the economic and political systems, with the exact position in part a function of development of his professional career. I have been principally concerned with examining, on the one hand, external relations of the scientific system and, on the other, with delineating external obligations of individual scientists, paying specific attention to relationships with the political system. If the most singular characteristics of modern science are to be fully understood, sociologists of science must discard their assumption of the autonomy of the social system of science.

Index